Selected Titles in This Series

634 **Darrin D. Frey,** Conjugacy of Alt_5 and $\text{SL}(2,5)$ subgroups of $E_8(\mathbb{C})$, 1998

633 **Dikran Dikranjan and Dmitri Shakhmatov,** Algebraic structure of pseudocompact groups, 1998

632 **Shouchuan Hu and Nikolaos S. Papageorgiou,** Time-dependent subdifferential evolution inclusions and optimal control, 1998

631 **Ronnie Lee, Steven H. Weintraub, and J. William Hoffman,** The Siegel modular variety of degree two and level four/Cohomology of the Siegel modular group of degree two and level four, 1998

630 **Florin Rădulescu,** The Γ-equivariant form of the Berezin quantization of the upper half plane, 1998

629 **Richard B. Sowers,** Short-time geometry of random heat kernels, 1998

628 **Christopher K. McCord, Kenneth R. Meyer, and Quidong Wang,** The integral manifolds of the three body problem, 1998

627 **Roland Speicher,** Combinatorial theory of the free product with amalgamation and operator-valued free probability theory, 1998

626 **Mikhail Borovoi,** Abelian Galois cohomology of reductive groups, 1998

625 **George Xian-Zhi Yuan,** The study of minimax inequalities and applications to economies and variational inequalities, 1998

624 **P. Deift and K. T-R McLaughlin,** A continuum limit of the Toda lattice, 1998

623 **S. A. Adeleke and Peter M. Neumann,** Relations related to betweenness: Their structure and automorphisms, 1998

622 **Luigi Fontana, Steven G. Krantz, and Marco M. Peloso,** Hodge theory in the Sobolev topology for the de Rham complex, 1998

621 **Gregory L. Cherlin,** The classification of countable homogeneous directed graphs and countable homogeneous n-tournaments, 1998

620 **Victor Guba and Mark Sapir,** Diagram groups, 1997

619 **Kazuyoshi Kiyohara,** Two classes of Riemannian manifolds whose geodesic flows are integrable, 1997

618 **Karl H. Hofmann and Wolfgang A. F. Ruppert,** Lie groups and subsemigroups with surjective exponential function, 1997

617 **Robin Hartshorne,** Families of curves in $\mathbb{P}^3$ and Zeuthen's problem, 1997

616 **Serguei G. Bobkov and Christian Houdré,** Some connections between isoperimetric and Sobolev-type inequalities, 1997

615 **Michael A. Dritschel and Hugo J. Woerdeman,** Model theory and linear extreme points in the numerical radius unit ball, 1997

614 **Richard Warren,** The structure of k-CS-transitive cycle-free partial orders, 1997

613 **D. L. Flannery,** The finite irreducible linear 2-groups of degree 4, 1997

612 **Joan Porti,** Torsion de Reidemeister pour les variétés hyperboliques, 1997

611 **D. Ginzburg, I. Piatetski-Shapiro, and S. Rallis,** L functions for the orthogonal group, 1997

610 **Mark Hovey, John H. Palmieri, and Neil P. Strickland,** Axiomatic stable homotopy theory, 1997

609 **Liviu I. Nicolaescu,** Generalized symplectic geometries and the index of families of elliptic problems, 1997

608 **Christina Q. He and Michel L. Lapidus,** Generalized Minkowski content, spectrum of fractal drums, fractal strings, and the Riemann zeta-functions, 1997

607 **Adele Zucchi,** Operators of class C_0 with spectra in multiply connected regions, 1997

606 **Moshé Flato, Jacques C. H. Simon, and Erik Taflin,** Asymptotic completeness, global existence and the infrared problem for the Maxwell-Dirac equations, 1997

(Continued in the back of this publication)

Conjugacy of Alt$_5$ and SL(2, 5) Subgroups of E$_8(\mathbb{C})$

Memoirs
of the
American Mathematical Society

Number 634

Conjugacy of Alt_5
and $SL(2, 5)$ Subgroups
of $E_8(\mathbb{C})$

Darrin D. Frey

May 1998 • Volume 133 • Number 634 (end of volume) • ISSN 0065-9266

American Mathematical Society
Providence, Rhode Island

1991 *Mathematics Subject Classification.* Primary 22E40, 20C33, 30B35, 20D06.

Library of Congress Cataloging-in-Publication Data
Frey, Darrin D., 1966–
  Conjugacy of Alt_5 and $SL(2,5)$ subgroups of $E_8(\mathbb{C})$ / Darrin D. Frey.
    p. cm. — (Memoirs of the American Mathematical Society, ISSN 0065-9266 ; no. 634)
  "Volume 133, number 634 (end of volume)."
  Includes bibliographical references.
  ISBN 0-8218-0778-1 (alk. paper)
  1. Lie groups. 2. Representations of groups. I. Title. II. Series.
QA3.A57 no. 634
[QA387]
510 s—dc21
[512′.55]   98-2682
                                                                                                   CIP

Memoirs of the American Mathematical Society

This journal is devoted entirely to research in pure and applied mathematics.

Subscription information. The 1998 subscription begins with volume 131 and consists of six mailings, each containing one or more numbers. Subscription prices for 1998 are $435 list, $348 institutional member. A late charge of 10% of the subscription price will be imposed on orders received from nonmembers after January 1 of the subscription year. Subscribers outside the United States and India must pay a postage surcharge of $30; subscribers in India must pay a postage surcharge of $43. Expedited delivery to destinations in North America $35; elsewhere $110. Each number may be ordered separately; *please specify number* when ordering an individual number. For prices and titles of recently released numbers, see the New Publications sections of the *Notices of the American Mathematical Society*.

Back number information. For back issues see the *AMS Catalog of Publications*.

Subscriptions and orders should be addressed to the American Mathematical Society, P. O. Box 5904, Boston, MA 02206-5904. *All orders must be accompanied by payment.* Other correspondence should be addressed to Box 6248, Providence, RI 02940-6248.

Copying and reprinting. Individual readers of this publication, and nonprofit libraries acting for them, are permitted to make fair use of the material, such as to copy a chapter for use in teaching or research. Permission is granted to quote brief passages from this publication in reviews, provided the customary acknowledgment of the source is given.

Republication, systematic copying, or multiple reproduction of any material in this publication (including abstracts) is permitted only under license from the American Mathematical Society. Requests for such permission should be addressed to the Assistant to the Publisher, American Mathematical Society, P. O. Box 6248, Providence, Rhode Island 02940-6248. Requests can also be made by e-mail to reprint-permission@ams.org.

Memoirs of the American Mathematical Society is published bimonthly (each volume consisting usually of more than one number) by the American Mathematical Society at 201 Charles Street, Providence, RI 02904-2294. Periodicals postage paid at Providence, RI. Postmaster: Send address changes to Memoirs, American Mathematical Society, P. O. Box 6248, Providence, RI 02940-6248.

© 1998 by the American Mathematical Society. All rights reserved.
This publication is indexed in *Science Citation Index*®, *SciSearch*®, *Research Alert*®, *CompuMath Citation Index*®, *Current Contents*®/*Physical, Chemical & Earth Sciences*.
Printed in the United States of America.

⊗ The paper used in this book is acid-free and falls within the guidelines established to ensure permanence and durability.
Visit the AMS home page at URL: http://www.ams.org/

10 9 8 7 6 5 4 3 2 1   03 02 01 00 99 98

Table of Contents

Introduction and Preliminaries 1

The Dihedral group of order 6 26

The Dihedral Group of order 10 33

The Alt$_5$ and SL(2, 5) fusion patterns in G, $\mathcal{A}$, Δ and Ω 45

Fusion patterns of Alt$_5$ and SL(2, 5) subgroups of H 73

Fusion patterns of Alt$_5$ subgroups of $\mathcal{E}$ 111

Conjugacy classes of Alt$_5$ subgroups of G 115

Conjugacy classes of SL(2, 5) subgroups of G 135

Appendix 155

Table of Notation and Definitions 160

References 161

Abstract

Let G be the complex Lie group of type E_8, and let Alt_5 denote the alternating group of degree 5. We provide a near classification of the Alt_5 and $SL(2, 5)$ subgroups of G up to conjugacy based on the conjugacy classes of the elements of the subgroups, and we give some restrictions on those Alt_5 and $SL(2, 5)$ subgroups of G for which the conjugacy question has not been answered.

1991 Mathematics Subject Classification: 22E40, 20C33, 20B35, 20D06

Key Words and Phrases: alternating group of degree 5, SL(2, 5), complex Lie group of type E_8, representation theory, finite groups, conjugacy.

Chapter 1
Introduction and Preliminaries

The classification of simple complex Lie groups implies that any simple complex Lie group is a central quotient of a classical group (that is, one of SL(n, $\mathbb{C}$), SO(n, $\mathbb{C}$) or Sp(n, $\mathbb{C}$)) or an exceptional group (that is, one of $G_2(\mathbb{C})$, $F_4(\mathbb{C})$, $E_6(\mathbb{C})$, $E_7(\mathbb{C})$ or $E_8(\mathbb{C})$). The classical groups are well understood, but much remains to be learned about the exceptional groups. In particular, from the character theory of finite groups we can say which finite simple and quasisimple groups embed in a given classical group and how many conjugacy classes of embeddings of a given finite group there are in that classical group. Until recently, this sort of information was not known in general for the exceptional complex Lie groups.

Recently, Cohen, Griess, Ryba, Wales and others have proved theorems classifying the finite simple and quasisimple subgroups which embed in the various exceptional complex Lie groups (see [CoGr '87], [CGL '93] and [GrRy '96] for E_7 and E_8, [CoWa '92] for F_4 and E_6 and [CoWa '83] and [Gr '94] for G_2). A few papers have also been written on conjugacy of finite subgroups of exceptional complex Lie groups (we mention, in particular, [CoWa '83], [Gr '94] for subgroups of G_2, [CGL '93] for PSL(2, 61) in E_8 and [GrRy '96] and [Serre '96] for $PGL_2(31)$ and $SL_2(32)$ in E_8) but for the most part, the problem of conjugacy of embeddings of finite simple and quasisimple groups in exceptional complex Lie groups is in the early stages. We should also mention that conjugacy results have been obtained by this author for Alt_5 and SL(2, 5) subgroups of E_6 and F_4 and will have been submitted for publication by the time this is published.

Received by the editor August 18, 1995.

It is our objective to classify the smallest of these subgroups, namely, the alternating group of degree five, and SL(2, 5), up to conjugacy. The problem of classifying the smaller finite subgroups of E_8 is somewhat different in nature from the problem of classifying large finite subgroups because the number of classes is much larger and the centralizers have much larger dimension for smaller groups. Our main organizing tool for this endeavor will be the idea of a fusion pattern which is simply a class function from one group to another (we define the notion of fusion pattern in more detail in Definition 4.1). We will show that out of 112 Alt_5 fusion patterns in E_8, only 19 are consistent with embeddings of Alt_5 into E_8 and for 17 of those, the fusion pattern is enough to determine conjugacy of the subgroups. For the other two, we are unable from this analysis to determine how many E_8-classes exist other than to say that there is at least one class for each. (Kay Magaard has since proved that the fusion pattern is enough to determine conjugacy for one of these). Corresponding to each conjugacy class of groups there are either one or two classes of embeddings depending on whether or not the elements of order 5 are rational (see Definition 1.17) yielding at least 31 classes of embeddings.

For SL(2, 5), we show that out of 1,850,240 fusion patterns, only 55 correspond to SL(2, 5) subgroups of E_8, and for 49 of these, the fusion pattern is enough to determine conjugacy of groups. Three fusion patterns correspond to exactly two classes of groups. For the other three fusion patterns we are unable from this analysis to determine how many E_8 classes exist other than to say at least one class existed for each. We do prove, however, that any SL(2, 5) subgroup of E_8 with one of two of these problem fusion patterns is forced to be conjugate to a subgroup of E_8 of type A_2E_6 (and, in fact, subsequent work shows that fusion pattern is enough to determine conjugacy), and that an SL(2, 5) subgroup of E_8 with the third problem fusion pattern is forced to be conjugate to a subgroup of E_8 of type E_7. Again, each of the conjugacy classes of groups affords either one or two conjugacy classes of embeddings depending on whether the elements of order 10 are rational in E_8 yielding 100 classes of embeddings from the fusion patterns affording one or two classes of subgroups and at least 3 from the other three classes.

To accomplish this classification, we make heavy use of the character theory of these two finite groups to determine the number of Alt_5 and SL(2, 5) conjugacy classes of subgroups in large classical subgroups of E_8. Then we find ways of forcing a subgroup with a given fusion pattern to be conjugate to a subgroup of one of the classical groups in E_8 for which we had already determined conjugacy relations. Finally, in cases where the classical subgroup of E_8 have multiple conjugacy classes of Alt_5 or SL(2, 5) subgroups with a given fusion pattern, we determine whether the classes in the classical subgroup fuse in E_8, perhaps by forcing the Alt_5 or SL(2, 5) into another classical subgroup of E_8.

This work is based on the author's doctoral dissertation under Robert L. Griess, Jr. at the University of Michigan and the author would like to express his appreciation to Professor Griess for his guidance. Our main theorems are Theorem 7.5 (p. 132) and Theorem 8.1 (p. 150). We recommend that a first time reader begin with Notation 1.12, Table 1.14 and Notation 1.15 and then skip to Chapter 4.

We begin with a few general results from group theory and then we define our notation.

Notation 1.1 We use the exponential notation for the action of an element of a group on a set, i.e., X^σ represents X acted on by σ.

Lemma 1.2 Suppose G is a group with the descending chain condition for normal subgroups, $\sigma \in \text{Aut}(G)$ and N is a normal subgroup of G such that $N^\sigma \leq N$. Then $N^\sigma = N$.

  Proof. Suppose $N^\sigma < N$. Then $N^{\sigma^{i+1}} < N^{\sigma^i}$ $\forall i = 1, 2, \ldots$, and N^σ is normal in G since σ is an automorphism. Hence we have an infinite decreasing chain $N > N^\sigma > N^{\sigma^2} > \ldots$ of normal subgroups contrary to our assumption. ∎

Lemma 1.3 Suppose $G = H_1 \times \ldots \times H_n$, where H_i is a nonabelian simple group for $i = 1, \ldots, n$. Let $\sigma \in \text{Aut}(G)$. Then σ permutes the set $S = \{H_i \mid i = 1, \ldots, n\}$.

  Proof. Let N be a minimal normal subgroup of G. Suppose N is not one of the elements of S. Now since H_i is normal $\forall i$ and N is normal, $H_i \cap N$ is a normal subgroup of H_i. Hence $H_i \cap N = \{1\}$ since N is not one of the elements of S. Let $\pi_i : G \to H_i$ be the projection map onto H_i. Now $[N, H_i] \leq N \cap H_i = 1$ $\forall i$, so $[N^{\pi_i}, H_i] = 1$ $\forall i$ and then $N^{\pi_i} \leq Z(H_i) = 1$ $\forall i$. That is, $N^{\pi_i} = 1$ $\forall i = 1, \ldots, n$ which implies $N = 1$. So the minimal normal subgroups of G are the elements of S, and σ therefore permutes the elements of S. ∎

Definition 1.4 Suppose $G = G_1 G_2 \ldots G_n$, where $G_i \leq G$ for $i = 1, \ldots, n$, and $[G_i, G_j] = 1$ for $i \neq j$. We call G a *central product* of the groups G_i and write $G = G_1 \circ \ldots \circ G_n$. It is not difficult to show that G is a quotient of $= G_1 \times \ldots \times G_n$.

Corollary 1.5 Suppose $G = H_1 \circ \ldots \circ H_n$, where H_i is a quasisimple group for $i = 1, \ldots, n$. Let $\sigma \in \text{Aut}(G)$. Then σ permutes the set $S = \{H_i \mid i = 1, \ldots, n\}$.

Proof. Let $Z = Z(G)$. Then $G/Z \cong H_1/Z(H_1) \times ... \times H_n/Z(H_n)$, and $H_i/Z(H_i)$ is nonabelian simple for $i = 1, ..., n$. Now σ induces an automorphism ρ on G/Z. By Lemma 1.3, ρ permutes the factors of G/Z. So mod Z, σ permutes the factors of G. But the image under σ of H_i is quasisimple, and contained in $H_j Z$ for some j. Now $(H_j Z)' = H_j$ so any quasisimple subgroup of $H_j Z$ is contained in H_j. Hence, $H_i^\sigma \leq H_j$. In fact, since $G = H_1 \circ ... \circ H_n = H_1^\sigma \circ ... \circ H_n^\sigma$, H_i^σ is normal in H_j. So $(H_i^\sigma)' = H_i^\sigma \Rightarrow H_i^\sigma \not\leq Z(H_j) \Rightarrow H_i^\sigma = H_j$. Hence, σ permutes the set S. ∎

Lemma 1.6 Suppose $G = H_1 \circ ... \circ H_n \circ T$ where H_i is a quasisimple group with finite center for $i = 1, ..., n$, and T is an abelian subgroup with no subgroups of finite index. Let $\sigma \in \text{Aut}(G)$. Then σ permutes the set $S = \{H_i \mid i = 1, ..., n\}$ and stabilizes T.

Proof. Let $Z = Z(G) = Z(H_1) \circ ... \circ Z(H_n) \circ T$. $G/Z \cong H_1/Z(H_1) \times ... \times H_n/Z(H_n)$ and $H_i/Z(H_i)$ is nonabelian simple. Now σ induces an automorphism ρ on G/Z. By Lemma 1.3, ρ permutes the factors of G/Z. So mod Z, σ permutes the factors of G. Now $H_1 \circ ... \circ H_n = G'$, so σ restricts to an automorphism of G'. But then by Lemma 1.5, σ permutes the set S. Now $T \leq Z(G)$ char G so $T^\sigma \leq Z(G) = \langle Z(H_i), T \mid i = 1, ...,n \rangle$. But $T^\sigma \cong T$ and therefore has no subgroups of finite index. Since $|Z(H_i)| < \infty \; \forall i$, $T^\sigma \leq T$. ∎

Definition 1.7 A group is *perfect* if $G' = G$ where G' is the commutator subgroup of G, and is *quasisimple* if, in addition, $G/Z(G)$ is simple.

Lemma 1.8 Suppose $G = H_1 \circ ... \circ H_n \circ T$ where H_i is a perfect group for $i = 1, ..., n$, and T is an abelian subgroup with no subgroups of finite index. Suppose L is perfect and $L \leq G$, then $L \leq H_1 \circ ... \circ H_n$.

Proof. Suppose $x \in L$. Then since L is perfect, $x \in L' \leq G' = H_1 \circ ... \circ H_n$. ∎

Lemma 1.9 The homomorphic image of a perfect group is perfect.

Proof. Let G be perfect, $\eta : G \to H$ a homomorphism. Let $h \in H$, and $g \in G$ such that $\eta(g) = h$. Since G is perfect, $g = [a_1, b_1] ... [a_n, b_n]$ where $a_i, b_i \in G$. So $h = [\eta(a_1), \eta(b_1)] ... [\eta(a_n), \eta(b_n)] \in H'$. So H is perfect. ∎

Lemma 1.10 Let $\eta : G \to H$ be a surjection. Then $H' = \eta(G')$.

Proof. Let $h \in H'$. Then $h = [x_1, y_1] ... [x_n, y_n]$ where x_i, y_i are elements of H. Let $a_i \in \eta^{-1}(x_i)$, $b_i \in \eta^{-1}(y_i)$ and let $g = [a_1, b_1] ... [a_n, b_n]$. Then $g \in G'$, and $\eta(g) = h$. So $H' \subseteq \eta(G')$.

Conversely, let $h \in \eta(G')$. Then $h = \eta([a_1, b_1] ... [a_n, b_n]) = \eta([a_1, b_1]) ... \eta([a_n, b_n]) = [\eta(a_1), \eta(b_1)] ... [\eta(a_n), \eta(b_n)]$ where $a_i, b_i \in G$. Hence $h \in H'$ and $\eta(G') \subseteq H'$. ∎

Lemma 1.11 Suppose $\eta : G \to H$ is a surjection and that H is a perfect, finite group. Then $H = \eta(G')$, and $|G'| \geq |H|$. Moreover, if $|\ker \eta|$ is prime, then G' is perfect.

Proof. Let $h \in H$. Then $h = [x_1, y_1] ... [x_n, y_n]$ where x_i, y_i are elements of H since H is perfect. Let $a_i \in \eta^{-1}(x_i)$, $b_i \in \eta^{-1}(y_i)$. Then $g = [a_1, b_1] ... [a_n, b_n]$ is an element of G', and $\eta(g) = h$. Hence $G' \cap \eta^{-1}(h)$ is not empty for any $h \in H$. So $H = \eta(G')$, and $|G'| \geq |H|$. If $|\ker \eta|$ is prime then G is finite and $|G'| = |H|$ or $|G|$. If $|G'| = |G|$, then $G' = G$ and $G = G'$ is perfect. If $|G'| = |H|$, then G' is isomorphic to H and therefore is perfect. ∎

Notation 1.12 Now let G be the complex Lie group of type E_8. Fix a set Π of fundamental roots and a Chevalley basis for G. Let $\mathcal{T}$ be the standard maximal torus, θ the Chevalley involution for G, and χ the adjoint character for G. When we speak of the trace in G of an element g ε G, we mean $\chi(g)$. We index the roots of the fundamental set Π as below with the extra node in the extended Dynkin diagram labeled with 0.

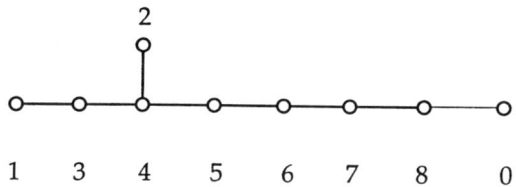

Let e_i (resp. e_{-i}) be the Chevalley generator of the root space for the fundamental root α_i (resp. $-\alpha_i$), and let $x_{\alpha_i}(t) := \exp(\operatorname{ad}(te_i))$ (resp. $x_{-\alpha_i} := \exp(\operatorname{ad}(te_{-i}))$) where $t \; \varepsilon \; \mathbb{C}$ (this is the notation used in [Carter '89]). Let X_i be the root subgroup of G generated by all elements of the form $x_{\alpha_i}(t)$, and X_{-i} be the root subgroup of G generated by all elements of the form $x_{-\alpha_i}(t)$. Let $S_i := \langle X_i, X_{-i} \rangle$. Finally, let S_0 be the fundamental $SL(2, \mathbb{C})$ subgroup of G corresponding to the extra node in the extended Dynkin diagram for G. Then we define the following standard subgroups of G.

Note 1.13 An elementary abelian p-group of order p^n is sometimes referred to as a p^n-group. So we refer to a fours-group, a nines-group and a sixteens-group and we mean elementary abelian groups of order 4, 9 and 16 respectively.

Table 1.14 Standard subgroups of G.

Notation	Meaning
$\mathcal{A}$	$\langle S_i \mid i \varepsilon \{0, 1, 3, 4, 5, 6, 7, 8\} \rangle$, a subgroup of G of type A_8
$\mathcal{S}$	$\langle S_i \mid i \varepsilon \{1, 2, 3, 4, 5, 6, 7\} \rangle$, a subgroup of G of type E_7

$\mathcal{E}$	$\langle S_i \mid i \in \{1, 2, 3, 4, 5, 6\}\rangle$, a subgroup of G of type E_6
$\mathcal{D}$	$\langle S_i \mid i \in \{2, 3, 4, 5\}\rangle$, a subgroup of G of type D_4
$\mathcal{G}$	By [Gr '91](2.18), there are two classes of outer automorphisms of $\mathcal{D}$ of order three with respective fixed point subgroups of type A_2 or G_2. We choose $\mathcal{G}$ to be the fixed point subgroup of type G_2 of a fixed outer automorphism of $\mathcal{D}$ of order 3 such that $\mathcal{G}$ contains $C(\mathcal{E})$.
$\mathcal{F}$	By [CoGr '87](3.9), the centralizer in G of $\mathcal{G}$ has type F_4. Let $\mathcal{F} = C(\mathcal{G})$.
H	$\langle S_i \mid i \in \{0, 2, 3, 4, 5, 6, 7, 8\}\rangle$, a subgroup of G of type D_8.
Δ	Let A be the centralizer of $\mathcal{E}$, and let y be the element of $\mathcal{E}$ of order 3 with $\mathcal{E}$-centralizer generated by $\langle S_i \mid i \in \{1, 2, 3, 5, 6\}\rangle$ and the fundamental $SL(2, \mathbb{C})$ subgroup of $\mathcal{E}$ corresponding to the extra node of the extended Dynkin diagram for $\mathcal{E}$. Then by [CoWa '92] (Table 2), $C_{\mathcal{E}}(y)$ has type $A_2A_2A_2$. Let $\Delta := A \circ C_{\mathcal{E}}(y)$. Then Δ is a subgroup of G of type A_2^4. It is the centralizer of a nines-group all of whose nontrivial elements have trace 5. One should note that by [Gr '91] (11.4) there are two conjugacy classes of nines-groups all of whose nontrivial elements have trace 5, namely the class containing the nines-group mentioned above, and a class whose members have connected centralizer of type D_4T_4.
Ω	Let y_1 be an element of $\mathcal{D}$ of order 2 with centralizer C_1 of type A_1^4, y_2 an element of $C(\mathcal{D})$ ($C(\mathcal{D})$ has type D_4 by

	[CoGr '87](3.7)) of order 2 with centralizer C_2 of type A_1^4. Let $\Omega := C_1 \circ C_2$. Then Ω has type A_1^8. Ω is the centralizer of the sixteens-group $Z(\Omega)$ of G.

Notation 1.15 We will be using the notation of [CoGr '87](Table 4) for classes of elements of order ≤ 6 in G, [CoGr '87](Table 6) for classes of elements of order ≤ 6 in S, and [CoWa '92](Table 2) for classes of elements of order ≤ 6 in $\mathcal{E}$. This notation as well as a list of the classes of elements of order 10 in G is presented in Tables 1.16, 1.19 and 1.22 which follow. We also adopt their notation for the alternating groups and the dihedral groups, namely, Alt_n for the alternating group of degree n, and Dih_n for the dihedral group of order n. The notation nX where X is a root system type and n is a positive integer denotes a perfect Lie group of type X over $\mathbb{C}$ with cyclic center of order n. The notation 2^2X denotes a perfect Lie group of type X over $\mathbb{C}$ with center isomorphic to $Z_2 \times Z_2$. We say such a group has type nX or type 2^2X. The isomorphism type of groups represented by notations nX and 2^2X will be determined by context. Finally, the notation X represents a perfect Lie group of type X but does not specify the order of the center.

Table 1.16 Conjugacy classes of elements of order ≤ 6 and 10 in G.

The information for elements of order ≤ 6 comes from [CoGr '87] (Table 4). The first column contains a label the first digit of which indicates the order k of this element; between square brackets are the exponents of nontrivial powers representing distinct conjugacy classes of elements of the same order and letters indicating the labels of conjugacy classes of powers by prime divisors of k. In the second column, s_0 is not included but can be deduced from $s_1, \ldots, s_8$ ([Kac '90] (Theorem 8.6) notation).

Class	$s = (s_1, s_2, \ldots, s_8)$ as in [Kac '90] (Theorem 8.6)	Multiplicity of $e^{2\pi i j/k}$ for $j = 0,\ldots,[(k+1)/2]$	Centralizer Type	Trace on the adjoint module for G.
2A	(0,0,0,0,0,0,0,1)	136, 112	A_1E_7	24
2B	(1,0,0,0,0,0,0,0)	120, 128	D_8	-8
3A	(0,1,0,0,0,0,0,0)	80, 84	A_8	-4
3B	(0,0,0,0,0,1,0)	86, 81	A_2E_6	5
3C	(1,0,0,0,0,0,0,0)	92, 78	D_7T_1	14
3D	(0,0,0,0,0,0,0,1)	134, 57	E_7T_1	77
4A[A]	(0,0,1,0,0,0,0,0)	66, 56, 70	A_7A_1	-4
4B[B]	(0,1,0,0,0,0,0,0)	64, 64, 56	A_7T_1	8
4C[B]	(0,0,0,0,1,0,0)	60, 64, 60	A_3D_5	0
4D[A]	(1,0,0,0,0,0,0,1)	70, 56, 66	$A_1D_6T_1$	4
4E[A]	(0,0,0,0,0,1,0)	82, 56, 54	$A_1E_6T_1$	28
4F[B]	(1,0,0,0,0,0,0,0)	92, 64, 28	D_7T_1	64
4G[A]	(0,0,0,0,0,0,0,1)	134, 56, 2	E_7T_1	132
5A[2]	(0,1,0,0,0,0,0,0)	64, 56, 36	A_7T_1	$8 + 20\tau$
5B[2]	(0,0,1,0,0,0,0,0)	52, 49, 49	$A_6A_1T_1$	3
5C	(0,0,0,0,1,0,0,0)	48, 50, 50	A_4A_4	-2
5D[2]	(1,0,0,0,0,1,0)	54, 51, 46	$A_2D_5T_1$	$3 + 5\tau$
5E[2]	(0,0,0,0,0,1,1)	82, 54, 29	$A_1E_6T_1$	$28 + 25\tau$
5F[2]	(2,0,0,0,0,0,0,0)	92, 64, 14	D_7T_1	$28 + 50\tau$
5G	(1,0,0,0,0,0,0,1)	68, 45, 45	D_6T_2	23
5H[2]	(0,0,0,0,0,0,0,2)	134, 56, 1	E_7T_1	$78 + 55\tau$
6A[B, A]	(0,1,0,0,0,0,0,0)	64, 56, 28, 16	A_7T_1	76
6B[B, D]	(1,0,1,0,0,0,0,0)	64, 29, 28, 70	A_7T_1	-5
6C[A, A]	(0,0,1,0,0,0,0,0)	54, 42, 42, 28	$A_6A_1T_1$	24
6D[B, C]	(0,1,0,0,0,0,0,1)	50, 43, 35, 42	A_6T_2	16
6E[A, C]	(1,1,0,0,0,0,0,0)	50, 35, 43, 42	A_6T_2	0
6F[A, B]	(0,0,1,0,0,0,0)	46, 36, 45, 40	$A_5A_2A_1$	-3
6G[A, A]	(0,1,0,0,0,1,0)	44, 38, 46, 36	$A_5A_2T_1$	0
6H[B, B]	(0,0,1,0,0,0,1)	42, 42, 39, 44	$A_5A_1A_1T_1$	1
6I[B, A]	(0,0,0,0,1,0,0,0)	40, 44, 40, 40	$A_4A_3T_1$	4
6J[B, C]	(1,0,0,0,0,1,0,0)	44, 40, 38, 48	$A_3D_4T_1$	-2
6K[B, B]	(0,0,0,0,0,1,0,0)	54, 48, 33, 32	$A_2D_5T_1$	37

6L[A, C]	(0,0,0,0,0,1,0,1)	52, 36, 42, 40	$A_1A_1D_5T_1$	6
6M[B, D]	(1,0,0,0,0,0,0,2)	70, 32, 25, 64	$A_1D_6T_1$	13
6N[A, D]	(2,0,0,0,0,0,0,1)	70, 24, 33, 64	$A_1D_6T_1$	-3
6O[A, B]	(1,0,0,0,0,0,1,0)	50, 38, 43, 36	$A_1D_5T_2$	9
6P[A, B]	(0,0,0,0,0,0,1,0)	82, 54, 27, 4	$A_1E_6T_1$	105
6Q[B, C]	(1,0,0,0,0,0,0,0)	92, 64, 14, 0	D_7T_1	142
6R[A, C]	(1,0,0,0,0,0,0,1)	68, 44, 34, 24	D_6T_2	54
6S[A, D]	(0,0,0,0,0,0,0,1)	134, 56, 1, 0	E_7T_1	189
6T[A, D]	(0,0,0,0,0,0,1,1)	80, 29, 28, 54	E_6T_2	27
10A[3] [A,B]	(0,1,0,0,0,0,0,0)	64,56,28,8,0,0	A_7T_1	$44 + 76\tau$
10B[3] [F,B]	(1,0,0,0,0,0,0,0)	92,64,14,0,0,0	A_7T_1	$78 + 78\tau$
10C[3] [H,B]	(3,0,1,0,0,0,0,0)	64,28,0,1,28,70	A_7T_1	$-5 - \tau$
10D[3] [B,A]	(0,0,1,0,0,0,0,0)	52,42,35,14,7,0	$A_6A_1T_1$	$31 + 56\tau$
10E[3] [A,A]	(2,0,1,0,0,0,0,0)	50,21,8,28,35,14	A_6T_2	$56 - 34\tau$
10F[3] [A,B]	(0,1,0,0,0,0,0,2)	50,35,14,22,21,14	A_6T_2	$44 + 6\tau$
10G[3] [F,B]	(0,1,0,0,0,0,0,3)	50,36,7,7,28,42	A_6T_2	$8 + 8\tau$
10H[3] [F,A]	(3,1,0,0,0,0,0,0)	50,28,7,7,36,42	A_6T_2	$8 - 8\tau$
10I[3] [B,B]	(0,1,0,0,0,0,0,1)	50,42,28,21,7,2	A_6T_2	$41 + 42\tau$
10J[3] [E,B]	(0,0,1,0,0,0,0,3)	42,30,15,14,24,40	$A_5A_1A_1T_1$	$1 + 7\tau$
10K[3] [E,A]	(2,0,0,1,0,0,0,0)	42,24,17,12,30,40	$A_5A_1A_1T_1$	$-3 - \tau$
10L[3] [A,A]	(0,1,0,0,0,0,1,2)	40,26,18,18,30,24	$A_5A_1T_2$	$16 - 4\tau$
10M[3] [E,B]	(2,0,1,0,0,0,0,1)	40,27,13,16,27,42	$A_5A_1T_2$	$1 - 3\tau$
10N[3] [A,A]	(0,0,1,0,0,0,0,2)	40,30,22,14,26,24	$A_5A_1T_2$	$8 + 12\tau$
10O[3] [B,B]	(0,0,1,0,0,0,0,1)	40,36,27,22,13,12	$A_5A_1T_2$	$23 + 28\tau$
10P[3] [B,A]	(0,1,0,0,0,0,1,0)	40,32,31,18,17,12	$A_5A_1T_2$	$15 + 28\tau$
10Q[3] [G,A]	(1,1,0,0,0,0,0,2)	38,22,26,19,23,30	A_5T_3	$1 + 6\tau$
10R[3] [B,A]	(1,1,0,0,0,0,0,1)	38,27,27,22,22,14	A_5T_3	$19 + 10\tau$
10S[3] [G,B]	(0,1,0,0,0,0,1,1)	38,27,23,22,18,30	A_5T_3	$7 + 10\tau$
10T[3] [C,B]	(0,0,0,0,1,0,0,0)	40,40,30,20,10,8	$A_4A_3T_1$	$22 + 40\tau$
10U[3] [C,A]	(0,0,0,1,0,0,0,0)	36,30,30,20,20,12	$A_4A_2A_1T_1$	$14 + 20\tau$
10V[3] [D,B]	(0,0,0,0,1,0,0,1)	34,33,25,21,18,20	$A_4A_2T_2$	$10 + 19\tau$
10W[3] [A,A]	(2,1,0,0,0,0,1,0)	34,25,20,16,31,30	$A_4A_2T_2$	-2τ
10X[3] [D,A]	(1,1,0,0,1,0,0,0)	34,21,21,25,30,20	$A_4A_2T_2$	$18 - 13\tau$
10Y[3] [A,B]	(0,0,0,0,1,0,0,2)	34,31,18,18,25,30	$A_4A_2T_2$	$4 + 6\tau$
10Z[3] [B,A]	(0,0,0,0,1,0,1,0)	32,24,27,22,25,20	$A_4A_1A_1T_2$	$7 + 4\tau$
10AA[3] [B,B]	(0,0,1,0,0,0,1,0)	32,30,25,24,19,20	$A_4A_1A_1T_2$	$11 + 12\tau$
10BB[3] [B,A]	(0,1,0,0,0,1,0,1)	30,23,27,22,26,22	$A_4A_1T_3$	$3 + 2\tau$

10CC[3] [D,B]	(0,0,1,0,0,0,1,1)	30,27,21,25,24,24	$A_4A_1T_3$	$10-\tau$
10DD[3] [D,A]	(1,0,1,0,0,0,1,0)	30,23,25,21,28,24	$A_4A_1T_3$	$2-\tau$
10EE[3] [A,B]	(2,0,0,0,1,0,0,0)	32,28,16,20,28,32	$A_3A_3T_2$	$4-4\tau$
10FF [C,B]	(1,0,0,0,1,0,0,0)	32,28,22,28,22,16	$A_3A_3T_2$	22
10GG [G,B]	(1,0,1,0,0,1,0,0)	32,23,22,23,22,36	$A_3A_3T_2$	-3
10HH[3] [D,A]	(0,0,0,1,0,0,0,2)	30,24,26,20,27,24	$A_3A_2A_1A_1T_1$	3τ
10II[3] [B,B]	(0,0,0,1,0,0,0,1)	28,28,25,24,21,24	$A_3A_2A_1T_2$	$3+8\tau$
10JJ[3] [B,A]	(0,0,1,0,1,0,0,0)	28,22,27,22,27,24	$A_3A_2A_1T_2$	-1
10KK[3] [C,A]	(0,0,1,0,0,1,0,0)	28,24,28,22,26,20	$A_3A_2A_1T_2$	$2+4\tau$
10LL[3] [B,B]	(1,0,0,0,1,0,0,1)	26,26,24,25,23,26	$A_3A_2T_3$	$1+2\tau$
10MM[3] [D,B]	(1,0,0,1,0,0,0,1)	26,26,22,24,25,28	$A_3A_1A_1A_1T_2$	$-\tau$
10NN[3] [F,B]	(3,0,0,0,0,1,0,0)	44,32,6,8,32,48	$A_3D_4T_1$	$-2-2\tau$
10OO [C,B]	(0,0,0,1,0,0,1,0)	24,26,24,26,24,24	$A_2A_2A_1A_1T_2$	2
10PP[3] [D,B]	(0,0,0,0,0,1,0,0)	54,48,30,16,3,0	$A_2D_5T_1$	$40+59\tau$
10QQ[3] [D,B]	(2,0,0,0,0,1,0,0)	38,24,14,32,27,16	$A_2D_4T_2$	$40-21\tau$
10RR[3] [F,A]	(0,0,0,0,0,1,0,3)	52,32,10,4,32,40	$A_1A_1D_5T_1$	$6+6\tau$
10SS [G,B]	(1,0,0,0,0,1,0,2)	36,24,21,24,21,32	$A_1A_1D_4T_2$	7
10TT [G,A]	(2,0,0,0,0,1,0,1)	36,20,25,20,25,32	$A_1A_1D_4T_2$	-1
10UU[3] [H,A]	(4,0,0,0,0,0,0,1)	70,24,1,0,32,64	$A_1D_6T_1$	$5-7\tau$
10VV[3] [H,B]	(1,0,0,0,0,0,0,4)	70,32,1,0,24,64	$A_1D_6T_1$	$5+9\tau$
10WW[3] [E,A]	(3,0,0,0,0,0,1,0)	50,22,11,18,32,32	$A_1D_5T_2$	$25-17\tau$
10XX[3] [E,B]	(0,0,0,0,0,1,0,2)	50,32,13,16,22,32	$A_1D_5T_2$	$21+7\tau$
10YY[3] [D,A]	(0,0,0,0,0,1,0,1)	50,34,26,20,17,4	$A_1D_5T_2$	$40+23\tau$
10ZZ[3] [D,A]	(1,0,0,0,0,0,2,1)	50,18,10,36,33,4	$A_1D_5T_2$	$72-41\tau$
10AAA[3] [D,A]	(1,0,0,0,0,1,0,1)	34,26,26,20,25,20	$A_1D_4T_3$	$8+7\tau$
10BBB[3] [E,A]	(0,0,0,0,0,0,1,0)	82,54,27,2,0,0	$A_1E_6T_1$	$57+79\tau$
10CCC[3] [H,A]	(0,0,0,0,0,0,0,1)	134,56,1,0,0,0	$A_1E_6T_1$	$133+57\tau$
10DDD[3] [G,A]	(1,0,0,0,0,0,0,1)	68,44,33,12,1,0	D_6T_2	$47+64\tau$
10EEE[3] [F,A]	(1,0,0,0,0,0,0,3)	68,32,2,12,32,24	D_6T_2	$54-10\tau$
10FFF [G,B]	(1,0,0,0,0,0,0,2)	68,32,13,32,13,0	D_6T_2	87
10GGG[3] [G,A]	(1,0,0,0,0,0,1,1)	48,27,26,19,18,20	D_5T_3	$21+16\tau$
10HHH[3] [E,A]	(1,0,0,0,0,0,1,2)	48,27,17,12,27,34	D_5T_3	$9+5\tau$
10III[3] [H,A]	(0,0,0,0,0,0,1,3)	80,28,0,1,28,54	E_6T_2	$27-\tau$
10JJJ[3] [E,A]	(0,0,0,0,0,0,1,2)	80,27,1,28,27,2	E_6T_2	$105-27\tau$

Proof. The elements of order 10 are obtained by finding all sequences $(s_0, s_1, \ldots, s_8)$ generating the unit ideal in the ring of integers and such that $\sum_{i=0}^{8} a_i s_i = 10$ where the a_i are the labels for $E_8^{(1)}$ in [Kac '90] (Table Aff 1). Then, by [Kac '90](Theorem 8.6), all elements of G of order 10 are exhausted up to conjugacy, and we can easily calculate the multiplicities of the eigenvalues of each class of elements. The centralizer types come from [Kac '90](Prop. 8.6(b)). ∎

Definition 1.17 An element of G of finite order is *rational* if it is conjugate to all of its powers.

Notation 1.18 By [CoGr '87](Section 4), the restriction to S of the adjoint module for G decomposes into a 3-dimensional fixed point space on which $C(S)$ acts adjointly, an adjoint module of dimension 133, and two irreducible modules of dimension 56. Let ξ be the adjoint character for S and κ the character afforded by the 56-dimensional modules.

Table 1.19 Conjugacy classes of elements of order ≤ 6 in S.

The information in the first, second, third and sixth columns on this Table is from [CoGr '87](Table 6). The fourth and fifth columns are the result of direct computation from the multiplicities on M (an irreducible 56-dimensional module), and the discussion at the beginning of section 4 in [CoGr '87].

Class	Centralizer in S	Multiplicity of $e^{2\pi ij/k}$ on M for $j = 0,1,...,[(k+1)/2]$	Trace on M	Trace on the adjoint module for S	Class in G
2A	E_7	0, 56	-56	133	2A
2B	A_1D_6	32, 24	8	5	2A
2C	A_1D_6	24, 32	-8	5	2B
3A	A_6T_1	14, 21	-7	7	3A
3B	E_6T_1	2, 27	-25	52	3B
3C	A_5A_2	20, 18	2	-2	3B
3D	$A_1D_5T_1$	20, 18	2	7	3C
3E	D_6T_1	32, 12	20	34	3D
4A[A]	A_7	0, 28, 0	0	-7	4A
4B[B]	$A_5A_1T_1$	12, 12, 20	-8	9	4A
4C[C]	A_5T_2	12, 16, 12	0	5	4B
4D[C]	$A_3A_3A_1$	12, 16, 12	0	-3	4C
4E[C]	$A_1D_5T_1$	4, 16, 20	-16	29	4C
4F[B]	$A_1A_1D_4T_1$	16, 12, 16	0	1	4D
4G[B]	D_6T_1	0, 12, 32	-32	65	4D
4H[A]	E_6T_1	0, 28, 0	0	25	4E
4I[B]	$A_5A_1T_1$	20, 12, 12	8	9	4E
4J[C]	$A_1D_5T_1$	20, 16, 4	16	29	4F
4K[B]	D_6T_1	32, 12, 0	32	65	4G
5A[2]	A_5T_2	12, 7, 15	$5 - 8\tau$	$11 + 4\tau$	5A
5B[2]	$A_4A_1T_2$	10, 12, 11	$-2 + \tau$	$4 - 2\tau$	5B
5C[2]	A_6T_1	0, 21, 7	$-21 + 14\tau$	$42 - 28\tau$	5B
5D[2]	$A_4A_2T_1$	6, 10, 15	$-4 - 5\tau$	$3 + 10\tau$	5C
5E[2]	$A_3A_2A_1$	12, 10, 12	$2 - 2\tau$	τ	5D
5F[2]	D_5T_2	2, 10, 17	$-8 - 7\tau$	$30 - 9\tau$	5D
5G[2]	$A_5A_1T_1$	20, 6, 12	$14 - 6\tau$	$9 + 13\tau$	5E
5H[2]	E_6T_1	0, 1, 27	$-1 - 26\tau$	$79 - 27\tau$	5E
5I[2]	$A_1D_5T_1$	20, 16, 2	$4 - 14\tau$	$17 + 22\tau$	5F
5J	$A_1D_4T_2$	16, 10, 10	6	8	5G
5K[2]	D_6T_1	32, 0, 12	$32 - 12\tau$	$35 + 31\tau$	5H
6A[C, A]	A_5T_2	12, 15, 6, 2	19	35	6A
6B[C, E]	A_5T_2	12, 6, 6, 20	-8	8	6B
6C[A, A]	A_6T_1	0, 21, 0, 14	7	7	6C
6D[B, A]	$A_4A_1T_2$	10, 10, 11, 4	5	11	6C
6E[C, D]	A_4T_3	10, 11, 7, 10	4	5	6D
6F[B, D]	A_4T_3	10, 7, 11, 10	-4	5	6E

6G[A, C]	A_5A_2	0, 18, 0, 20	-2	-2	6F
6H[B, B]	$A_5A_1T_1$	2, 12, 15, 0	-1	-4	6F
6I[B, C]	$A_3A_2A_1T_1$	8, 6, 12, 12	-10	14	6F
6J[B, A]	A_5T_2	2, 6, 15, 12	-19	35	6G
6K[B, A]	$A_3A_2T_2$	8, 9, 12, 6	-1	-1	6G
6L[C, C]	$A_5A_1T_1$	0, 6, 12, 20	-26	50	6H
6M[C, B]	$A_5A_1T_1$	0, 15, 12, 2	1	-4	6H
6N[C, C]	$A_3A_1A_1T_2$	8, 10, 8, 12	-2	2	6H
6O[C, A]	$A_4A_1T_2$	4, 11, 10, 10	-5	11	6I
6P[C, A]	$A_3A_2T_2$	6, 12, 9, 8	1	-1	6I
6Q[C, D]	$A_3A_1A_1A_1T_2$	8, 10, 8, 12	-2	-1	6J
6R[C, D]	$A_1D_4T_2$	4, 8, 10, 16	-14	23	6J
6S[C, C]	$A_3A_2A_1T_1$	12, 12, 6, 8	10	14	6K
6T[C, B]	D_5T_2	2, 16, 11, 0	7	20	6K
6U[B, D]	$A_3A_1A_1A_1T_1$	12, 8, 10, 8	2	-1	6L
6V[A, D]	$A_1D_5T_1$	0, 18, 0, 20	-2	7	6L
6W[B, D]	$A_1D_5T_1$	0, 2, 16, 20	-34	71	6L
6X[C, E]	$A_1A_1D_4T_1$	16, 8, 4, 16	4	2	6M
6Y[C, E]	D_6T_1	0, 0, 12, 32	-44	98	6M
6Z[B, E]	$A_1A_1D_4T_1$	16, 4, 8, 16	-4	2	6N
6AA[A, E]	D_6T_1	0, 12, 0, 32	-20	34	6N
6BB[B, C]	$A_3A_1A_1T_2$	12, 8, 10, 8	2	2	6O
6CC[B, B]	D_5T_2	0, 11, 16, 2	-7	20	6O
6DD[B, C]	$A_5A_1T_1$	20, 12, 6, 0	26	50	6P
6EE[A, B]	E_6T_1	0, 27, 0, 2	25	52	6P
6FF[C, D]	$A_1D_5T_1$	20, 16, 2, 0	34	71	6Q
6GG[B, D]	$A_1D_4T_2$	16, 10, 8, 4	14	23	6R
6HH[B, E]	D_6T_1	32, 12, 0, 0	44	98	6S
6II[B, E]	A_5T_2	20, 6, 6, 12	8	8	6T

Table 1.20 Selected classes of elements of order 10 in S.

The labels represent the G-class which these S-classes are contained in. The letters in the brackets denote the S-class of the square and fifth power respectively. The multiplicities were calculated as described in Table 5.13, except that inner products were taken only with roots from E_7.

Class in G	Multiplicity of $e^{2\pi i/k}$ for j = 0, ..., 5 on the adjoint module for S	$\xi(g)$	$\kappa(g)$
10E[A, B]	27, 11, 6, 16, 15, 10	$27 - 14\tau$	$13 - 10\tau$
10L[A, B]	21, 12, 14, 12, 10, 16	$3 + 4\tau$	$5 - 4\tau$
10L[A, B]	37, 6, 0, 26, 16, 0	$63 - 36\tau$	$-25 + 16\tau$
10N[A, B]	21, 10, 12, 14, 12, 16	$7 - 4\tau$	$-1 + 8\tau$
10N[A, A]	37, 0, 22, 0, 26, 0	$15 - 4\tau$	$-5 + 8\tau$
10P[B, B]	21, 12, 9, 16, 15, 8	$20 - 10\tau$	$-4 + 19\tau$
10Q[J, B]	19, 12, 12, 13, 13, 14	$6 - 2\tau$	$-1 - 2\tau$
10R[B, B]	19, 15, 13, 12, 12, 10	$8 + 4\tau$	$4 + 3\tau$
10X[E, B]	19, 15, 13, 13, 12, 8	$11 + 3\tau$	$2 - 8\tau$
10X[F, B]	27, 7, 11, 15, 10, 20	$11 - 7\tau$	$2 - 3\tau$
10Z[B, B]	17, 14, 13, 12, 13, 12	$4 + 2\tau$	τ
10Z[C, B]	29, 2, 15, 20, 5, 20	$14 - 8\tau$	$-5 + 6\tau$
10Z[B, A]	29, 0, 25, 0, 27, 0	$4 - 2\tau$	$2 - \tau$
10BB[B, B]	15, 13, 13, 12, 14, 14	0	τ
10BB[B, B]	27, 11, 5, 20, 16, 2	$40 - 20\tau$	$-20 + 11\tau$
10DD[E, B]	15, 13, 13, 13, 14, 12	$3 - \tau$	-2
10DD[F, B]	27, 11, 15, 11, 6, 20	$3 + 9\tau$	$-2 - 5\tau$
10HH[F, B]	23, 8, 14, 12, 9, 24	$-3 + \tau$	τ
10HH[E, B]	27, 24, 18, 8, 3, 0	$17 + 31\tau$	$-10 - 14\tau$
10HH[E, B]	19, 12, 10, 16, 15, 8	$17 - 9\tau$	$-10 + 6\tau$
10HH[E, A]	27, 0, 27, 0, 26, 0	τ	-2τ
10JJ[B, B]	17, 14, 15, 12, 11, 12	$2 + 6\tau$	$-3 - 3\tau$
10JJ[C, B]	25, 16, 3, 4, 19, 24	$2 - 4\tau$	$-3 + 2\tau$
10JJ[B, B]	21, 10, 9, 18, 15, 8	$22 - 14\tau$	$-13 + 7\tau$
10KK[D, B]	17, 14, 10, 10, 16, 16	$1 - 2\tau$	$-1 + 3\tau$
10KK[D, B]	21, 10, 14, 16, 10, 12	$11 - 2\tau$	$-6 + 3\tau$
10KK[D, B]	25, 12, 4, 16, 18, 8	$29 - 18\tau$	$-15 + 11\tau$
10AAA[E, B]	15, 12, 13, 14, 14, 12	$4 - 3\tau$	$2 + 2\tau$
10AAA[F, B]	31, 8, 10, 16, 9, 16	$21 - 7\tau$	$-8 + 7\tau$
10EEE[C, B]	33, 16, 2, 8, 16, 16	$23 - 6\tau$	$14 - 2\tau$

Notation 1.21 The restriction to $\mathcal{E}$ of the adjoint module for G decomposes into an 8-dimensional fixed point space on which $C(\mathcal{E})$ acts adjointly, an adjoint module adj($\mathcal{E}$) of dimension 78, three irreducible modules of dimension 27 and their duals.

Table 1.22 Conjugacy classes of real elements of order ≤ 6 in $\mathcal{E}$.

The information in the first, second, third and fifth columns on this Table is from [CoWa '92](Table 2). The fourth, sixth and seventh columns are the result of direct computation from the multiplicities on K and the adjoint module for $\mathcal{E}$, adj($\mathcal{E}$).

Class	Centralizer in $\mathcal{E}$	Multiplicity of $e^{2\pi i j/k}$ on K for $j = 0, 1, ..., [(k+1)/2]$	Trace on K	Multiplicity of $e^{2\pi i j/k}$ on adj($\mathcal{E}$) for $j = 0, 1, ..., [(k+1)/2]$	Trace on adj($\mathcal{E}$)	Class in G
2A	A_5A_1	15, 12	3	38, 40	-2	2A
2B	D_5T_1	11, 16	-5	46, 32	14	2B
3A	A_5T_1	15, 6	9	36, 21	15	3D
3C	$A_2A_2A_2$	9, 9	0	24, 27	-3	3B
3D	D_4T_2	9, 9	0	30, 24	6	3C
4A[A]	A_5T_1	15, 6, 0	15	36, 20, 2	34	4G
4D[B]	$A_3A_1A_1T_1$	5, 8, 6	-1	22, 16, 24	-2	4C
4E[A]	$A_3A_1T_2$	7, 6, 8	-1	20, 20, 18	2	4D
4F[A]	$A_2A_2A_1T_1$	9, 6, 6	3	20, 20, 18	2	4E
4H[B]	D_4T_2	9, 8, 2	7	30, 16, 16	14	4F
5A[2]	A_5T_1	15, 6, 0	$9 + 6\tau$	36, 20, 1	$16 + 19\tau$	5H
5E	A_3T_3	7, 5, 5	2	18, 15, 15	3	5G
5F[2]	$A_2A_2A_1T_1$	9, 3, 6	$6 - 3\tau$	20, 11, 18	$9 - 7\tau$	5E
5G[2]	$A_2A_1A_1T_2$	5, 5, 6	$-\tau$	16, 16, 15	τ	5D
5I[2]	D_4T_2	9, 1, 8	$8 - 7\tau$	30, 8, 16	$22 - 8\tau$	5F
6A[B, D]	A_3T_3	3, 5, 4, 6	-2	18, 10, 14, 12	2	6J
6B[A, D]	$A_1A_1A_1A_1T_2$	5, 4, 5, 4	0	14, 12, 12, 16	-2	6L

6C[A, C]	$A_1A_1A_1T_3$	5, 4, 5, 4	0	12, 14, 13, 12	1	6O
6D[B, C]	$A_2A_1A_1T_2$	5, 6, 3, 4	4	16, 12, 15, 8	5	6K
6E[B, A]	$A_3A_1T_2$	7, 4, 2, 8	1	20, 8, 13, 16	-1	6M
6F[A, A]	$A_3A_1T_2$	7, 2, 4, 8	-3	20, 12, 9, 16	7	6N
6G[A, D]	A_3T_3	7, 5, 4, 2	6	18, 14, 10, 12	10	6R
6H[A, C]	$A_2A_2A_1T_1$	9, 6, 3, 0	12	20, 18, 9, 4	25	6P
6I[A, A]	$A_2A_2T_2$	9, 3, 3, 6	3	18, 11, 10, 18	1	6T
6J[B, D]	D_4T_2	9, 8, 1, 0	16	30, 16, 8, 0	38	6Q
6K[A, A]	A_5T_1	15, 6, 0, 0	21	36, 20, 1, 0	55	6S

Remark 1.23 Having mentioned [CoGr '87] and [CoWa '92] we point out some errors and/or unclear statements in those papers. In [Co Gr '87](Table 1), the unique fixed point free character for L(3, 5) should be $124_d + 124_e$. In Table 4 of [CoGr '87] the cube of an element of type 6J should have type 2B and the cube of an element of type 6L should have type 2A. In Table 6 of [CoGr '87], the cube of an element of type 6V should have type 2A, and the centralizer of an element of type 3E should have type D_6T_1 (this was incorrectly corrected in [Gr '91] in the paragraph before (2.14)). These changes are made in the Tables above.

In [CoWa '92](Lemma 8.2), it is not clear whether the authors intended to say that all SL(2, 5)-subgroups of $\mathcal{E}$ with central involution of type 2A occur in subgroups of $\mathcal{E}$ isomorphic to SL(2, $\mathbb{C}$), but this is certainly not true. For example, Table 5.36 (cases 16, 17, 18) shows that there are SL(2, 5)-subgroups of $\mathcal{D} \subseteq \mathcal{E}$ which are not contained in any SL(2, $\mathbb{C}$)-subgroup of $\mathcal{E}$.

Lemma 1.24 There is a unique conjugacy class of each of the following types of subgroup in G: $2E_7, 3E_6, 2D_8, 4D_7$ and $3A_8$.

Proof. Let X be a subgroup of G of one of the above types. Suppose Z(X) = <x>. Then $X \leq C(x)$ in such a way that $Z(X) \leq Z(C(x))$. Now C(x) is the centralizer of an element of order |x| and so must have one of the types listed for elements of order $|x|$ in Table 1.16. But in each of the cases listed in the statement, only one of the possible centralizers has X as a subgroup such that $Z(X) \leq Z(C(x))$. Also, in each case, X is the unique subgroup of $C(x)$ of its type. Hence if Y is any other subgroup of G which has the same type as X, then $Z(Y)$ is conjugate to $Z(X)$, say $Z(Y)^g = Z(X)$, and then Y^g is a subgroup of $C(x)$ with the same type as X which forces $Y^g = X$. ∎

Lemma 1.25 There is a unique conjugacy class of subgroups of G of type 2^2D_6.

Proof. Let X be a subgroup of G of type 2^2D_6. Then $Z(X)$ is a fours-group, and $C(Z(X))$ must contain X. Hence by [CoGr '87](Table 5), $Z(X)$ has type BAA, so $C(Z(X))$ has type $A_1A_1D_6$. Suppose Y is another subgroup of G of type 2^2D_6. Then $Z(Y)$ is a fours-group of type BAA so, by [CoGr '87](3.7), $\exists g \in G$ such that $Z(Y)^g = Z(X)$, and then $Y^g \leq C(Z(X))$. But $C(Z(X))$ has only one subgroup of type 2^2D_6 and so $Y^g = X$. ∎

Lemma 1.26 There are three G-classes of subgroups of G of type $6A_5$, at least two of which have representatives in $\mathcal{A}$. Moreover, there are two $\mathcal{E}$-classes of subgroups of $\mathcal{E}$ of type $6A_5$ both of which have representatives in $\mathcal{A}$.

Proof. Let X be a subgroup of G of type $6A_5$. Then $Z(X)$ is generated by an element x of order 6. Now $C(x)$ contains X in such a way that $Z(X) \leq Z(C(x))$. By Table 1.16, x has type 6F, 6G or 6H. If x has type 6F, then $C(x)$ contains a group of type $SL(6, \mathbb{C})$ since the centers of the other factors of $C(x)$ do not have elements of order 6. If x has type 6G or 6H, then the factor of $C(x)$ of type A_5 is generated by fundamental $SL(2, \mathbb{C})$'s so is isomorphic to $SL(6, \mathbb{C})$. If x has type 6F or 6G, then X is conjugate to a subgroup of $\mathcal{A}$ because its generating subgroups are subgroups of $\mathcal{A}$. Next, suppose that X is a subgroup of $\mathcal{E}$ which has type $6A_5$. Then $Z(X)$ is generated by an element x of $\mathcal{E}$ of order

6. By [CoWa '92](Table 2), x has type $6K[\mathcal{E}]$, $6L[\mathcal{E}]$ or $6M[\mathcal{E}]$. Now $6K[\mathcal{E}] \subseteq 6S[G]$, $6L[\mathcal{E}] \subseteq 6G[G]$ and $6M[\mathcal{E}] \subseteq 6F[G]$. Clearly, if x has type $6L[\mathcal{E}]$ or $6M[\mathcal{E}]$, then X is conjugate to a subgroup of $\mathcal{A}$. Finally, x cannot have type 6K since $C_G(x)$ doesn't contain X in such a way that $Z(X) \leq Z(C_G(x))$. ∎

Lemma 1.27 There is a unique $\mathcal{E}$-class of subgroups of $\mathcal{E}$ of type $4D_5$.

Proof. Let X be a subgroup of $\mathcal{E}$ of type $4D_5$. Then $Z(X)$ is generated by an element x of order 4. Now $C(x)$ contains X in such a way that $Z(X) \leq Z(C(x))$. By [CoWa '92](Table 2), x has type $4G[\mathcal{E}]$ or $4G[3][\mathcal{E}]$, and then x^{-1} has the opposite type. Hence every subgroup of $\mathcal{E}$ of type $4D_5$ is conjugate to X. ∎

Definition 1.28 Let $G = G_1 \circ ... \circ G_n$ be a central product and define $G^i := <G_j \mid j \neq i>$, $Z_i := G_i \cap G^i$ and $Z := <Z_i \mid i = 1, ..., n> \leq Z(G)$. There is a natural map $G \to \Pi G_i/Z_i$, defined by $(g_i) \mapsto (Z_i g_i)$ (easily, one sees this is well-defined) with kernel Z. Let S be a subgroup of G. The *quasiprojections* of S are the groups $S_i \leq G$, $i = 1, ..., n$, which satisfy: $Z_i \leq S_i$ and S_i/Z_i is the projection of SZ/Z into the ith factor with respect to the decomposition $\Pi G_i/Z_i$. This definition comes from [Gr '91](2.5).

Lemma 1.29 There is a unique G-class of subgroups of G of type 2^2D_4 with center of type BBB, and connected centralizer of type 2^2D_4.

Proof. By [CoGr '87](3.7), any fours-group of type BBB is conjugate to the center of $D := \mathcal{D} \circ C_G(\mathcal{D})$. Hence, any subgroup S of G of type 2^2D_4 with center of type BBB and connected centralizer of type 2^2D_4 is conjugate to a subgroup of D. Suppose S is a diagonal subgroup of D. Let S_1 be the quasiprojection (see Definition 1.28) of S into $\mathcal{D}$ in the sense of [Gr '91](2.5), and S_2 the quasiprojection of S into $C(\mathcal{D})$. Then S_1 is a subgroup of $\mathcal{D}$ of type

D_4 and S_2 is a subgroup of $C(\mathcal{D})$ of type D_4 so $S_1 = \mathcal{D}$, and $S_2 = C(\mathcal{D})$. Suppose $t = t_1 t_2$ is an element of $C_D(S)$ where $t_1 \in \mathcal{D}$ and $t_2 \in C(\mathcal{D})$. Let s be any element of S, say $s = s_1 s_2$ where $s_1 \in S_1$, $s_2 \in S_2$. Then $t_1 t_2 s_1 s_2 = t_1 s_1 t_2 s_2 = s_1 s_2 t_1^{(s_1)} t_2^{(s_2)} = s_1 s_2 t_1 t_2$ since $t \in C_D(S)$, so $t_i^{(s_i)} \in t_i Z(\mathcal{D})$ for $i = 1, 2$. Since s_1 is arbitrary in $\mathcal{D}$, $t_1 \in B$ where $B := \{b \in \mathcal{D} \mid [b, \mathcal{D}] \subseteq Z(\mathcal{D})\}$. Now B is a group since if $b_1, b_2 \in B$, $d \in \mathcal{D}$, $[b_1, d] = z_1$ and $[b_2, d] = z_2$, then we have $[b_1 b_2, d] = b_2^{-1}(b_1^{-1} d^{-1} b_1) b_2 d = b_2^{-1} z_1 d^{-1} b_2 d = z_1 [b_2, d] = z_1 z_2 \in Z(\mathcal{D})$.

Define $\phi : B \to \{\text{subgroups of } Z(\mathcal{D})\}$ by $\phi(b) := [b, \mathcal{D}]$. Then ϕ is a homomorphism since $\phi(b_1 b_2) = [b_1 b_2, \mathcal{D}] = b_2^{-1} b_1^{-1} \mathcal{D} b_1 b_2 \mathcal{D} = b_2^{-1} \phi(b_1) \mathcal{D} b_2 \mathcal{D} = \phi(b_1) \phi(b_2)$. Now $\ker \phi = Z(\mathcal{D})$ so since $\{\text{subgroups of } Z(\mathcal{D})\}$ is finite, B is finite and therefore $C(S)$ is finite, contradicting our assumption. Hence S is not a diagonal subgroup of D and is therefore conjugate to one of the factors of $\mathcal{D} \circ C_G(\mathcal{D})$. But, by [CoGr '87](3.7), $\mathcal{D}$ and $C_G(\mathcal{D})$ are conjugate in G and we have the result. ∎

Lemma 1.30 The Chevalley involution θ inverts $\mathcal{T}$.

Proof. If $h = h(\chi) \in \mathcal{T}$, and α is a root, then $\chi(-\alpha) = \chi(\alpha^{-1})$, i.e. $\chi^\theta = \chi^{-1}$ so $h(\chi)^\theta = h(\chi^{-1}) = h(\chi)^{-1}$ so θ inverts $\mathcal{T}$. (Here we are using the notation of [Cart '89]). ∎

Corollary 1.31 If t, t' are elements of a torus $T \leq \mathcal{T}$, then $t\theta \sim_T t'\theta$

Proof. If we take $s \in T$ such that $s^2 = t' t^{-1}$ then $s(t\theta) s^{-1} = s^2 t\theta = t' t^{-1} t \theta = t'\theta$.

Lemma 1.32 Let K be a semisimple subgroup of G generated by root groups with root lattice $M \leq L$. If $\dim M < 8$ and $\det M < 8$, then K is simply connected.

Proof. It is enough to show that M is a direct summand of L by [Gr '91] (2.13)(iv). Suppose M is not a direct summand of L. Then $M \leq M_1 \leq L$ where M_1 is a direct summand of L with the same rank as M with $[M_1 : M] = d \in Z$. Now $\det M_1 = (1/d^2)(\det M)$ and $\det M_1 \in Z$. If no perfect square $\neq 1$ divides det M then $d = 1$ so we are done for det M = 1,2,3,5,6,7. Suppose det M = 4. Then $d > 1 \Rightarrow d = 2$. But then $\det M_1 = 1$ so M_1 is an even unimodular lattice which implies that $\dim M_1 \equiv 0 \pmod 8$, a contradiction. Hence $d = 1$ and M is a direct summand of L. ∎

Lemma 1.33 Let $y \in G$ such that $C_G(y)$ is a central product, say $C_G(y) = H_1 \circ ... \circ H_n$ where each H_i is either quasisimple or a torus. Denote $C_G(y)$ by C and $N_G(<y>)$ by N, and let $x \in N \setminus C$. Let $Y \leq Z(H_1) \circ ... \circ Z(H_n)$, h_i, h_i', $g_i \in H_i$ and $(h_i x)^{g_i} \equiv h_i' x \pmod{Y \cap H_i} \, \forall i$. Then $(h_1 h_2 ... h_n x)^{g_1 g_2 \cdots g_n} \equiv h_1' h_2' ... h_n' x \pmod Y$.

Proof. Write $g = g_1 ... g_n$. Then
$$(h_1 h_2 ... h_n x)^g = (h_1 ... h_{n-1})^g (h_n x)^{(g_n g_1 \cdots g_{n-1})}$$
$$\equiv (h_1 ... h_{n-1})^g (h_n' x)^{g_1 \cdots g_{n-1}} \pmod{Y \cap H_n}$$
$$\equiv (h_1 ... h_{n-1})^{g_1 \cdots g_{n-1}} (h_n') x^{g_1 \cdots g_{n-1}} \pmod{Y \cap H_n}$$
$$\equiv h_n' (h_1 ... h_{n-1} x)^{g_1 \cdots g_{n-1}} \pmod{Y \cap H_n}.$$

By induction,
$$h_n'(h_1 ... h_{n-1} x)^{g_1 \cdots g_{n-1}} \equiv h_n'(h_1' ... h_{n-1}' x) \pmod{Y \cap (H_1 \circ ... \circ H_{n-1})}$$
$$\equiv (h_1' ... h_n' x) \pmod{Y \cap (H_1 \circ ... \circ H_{n-1})}.$$
Hence $(h_1 h_2 ... h_n x)^{g_1 \cdots g_n} \equiv h_1' h_2' ... h_n' x \pmod Y$. ∎

Lemma 1.34 Suppose H is a group, K a normal subgroup of H, and $H = K<t>$ for some $t \in H$. Then K controls fusion in Kt, i.e. if $x, y \in Kt$ and $x \sim_H y$ then $x \sim_K y$.

Proof. Say $y = x^h$, where $h \in H$. Now $H = K\langle t \rangle = K\langle v \rangle = \langle v \rangle K$ for any $v \in Kt = tK$. $x \in Kt$ so $h = wk$ for some $w \in \langle x \rangle$, $k \in K$. So $y = x^h = x^{wk} = x^k$. ∎

Lemma 1.35 Suppose $y \in 5A$ or $5B$. Then $C_G(y)$ is a central product containing X and T as factors where X is a subgroup of G of type A_n generated by subgroups S_i as in Notation 1.12, with $n \leq 7$, T is a torus and $X \cap T = Z(X)$.

Proof. By Table 1.16, we need only prove that $X \cap T = Z(X)$. Now X is contained in a fundamental A_{n+1}-subgroup A of G generated by subgroups S_i as in Notation 1.12. In fact XT corresponds to a subgroup B of A of the form $\left\{ \begin{bmatrix} M & 0 \\ 0 & c \end{bmatrix} \mid M \in GL_{n+1}, c \in \mathbb{C}^\times, (\det M)c = 1 \right\}$. Now $X \leftrightarrow \left\{ \begin{bmatrix} M & 0 \\ 0 & 1 \end{bmatrix} \mid M \in SL_{n+1} \right\} \leq B$, $T \leftrightarrow \left\{ \begin{bmatrix} \lambda I_{n+1} & 0 \\ 0 & c \end{bmatrix} \mid \lambda^{n+1} c = 1 \right\} \leq B$. A matrix $\begin{bmatrix} M & 0 \\ 0 & c \end{bmatrix} \in X \cap T$ if and only if M is of the form λI_{n+1} with $\lambda^{n+1} = 1$ and $c = 1$. So $X \cap T \cong Z_{n+1} \cong Z(X)$, i.e. $X \cap T = Z(X)$. ∎

Lemma 1.36 If S is a subgroup of G isomorphic to Alt_4 all of whose involutions have type $2A$, then $C_G(S)^o$ has type $A_2A_2A_2$, D_4T_2, A_5T_1, or E_6 according as the trace of an element of order 3 of S equals -4, 5, 14 or 77.

Proof. Let W be the fours group in S. By [CoGr '87] (3.7), $C := C_G(W)^o$ has type E_6T_2. Since the center of the E_6 factor of C has order 3, W is contained in the T_2 part of C.

Let y be an element of order 3 in S. Then the action of y on T_2 does not induce an inner automorphism of T_2 since y permutes the involutions of W. But y does induce an inner automorphism on the E_6 factor since a group of type E_6 has no outer automorphisms of order 3. So y induces an inner automorphism determined by an element of order 3 or an element of order 9 whose cube is in the center of the E_6 factor of C. The possibilities are listed in [Gr '91](Table 6) and the dimensions are 24, 30, 36 and 78 by the orthogonality relations. The only centralizer types which satisfy those dimension requirements are the ones listed in the statement. ∎

Lemma 1.37 Suppose y is an element of $\mathcal{T}$ of finite order such that $C := C_G(y)$ is of the form $H_1 \circ \ldots \circ H_n \circ T$ where H_i is quasisimple and simply connected, H_i and H_j have different types for $i \neq j$, $H_i \cap T = Z(H_i)$ for $i = 1, \ldots, n$, and T is a torus contained in $\mathcal{T}$. Assume θ is outer on H_i for $i = 1, \ldots, n$. Let c_i be the number of conjugacy classes of outer automorphisms of $H_i/Z(H_i)$ of order 2 in the outer automorphism class of θ for each i. Then the number of C-classes of elements of $C\theta$ which induce outer automorphisms of C of order two is equal to $c_1 \ldots c_n$.

Proof. For each $i \in \{1, \ldots, n\}$, since the elements of $H_i\theta$ induce automorphisms of H_i, there are c_i H_i-classes of elements of $H_i\theta$ which induce outer automorphisms of order two. Since $y \in Z(C) = Z(H_1) \circ \ldots \circ Z(H_n) \circ T = T$ (since $Z(H_i) \leq T$ for all i), $c\theta$ inverts y, for $c \in C$, and hence $c\theta \in N(\langle y \rangle) \backslash C$.

Want to show: if $w_i, w_i' \in H_i$, for $i = 1, \ldots, n$, $t, t' \in T$, $Y = Z_1 \circ \ldots \circ Z_n \cap T$ (where $Z_i := Z(H_i)$), then $w_1 w_2 \ldots w_n t\theta \sim_C w_1' w_2' \ldots w_n' t'\theta \pmod{Y} \Leftrightarrow w_i\theta \sim_{H_i} w_i'\theta \pmod{Y \cap H_i}$ for $i = 1, \ldots, n$.

We have ($\Leftarrow$) by Lemma 1.33. Suppose $w_1 w_2 \ldots w_n t\theta \sim_C w_1' w_2' \ldots w_n' t'\theta \pmod{Y}$ say $(w_1 w_2 \ldots w_n t\theta)^{g_1 g_2 \ldots g_n s} = w_1' w_2' \ldots w_n' t'\theta x$ where $x \in Y$, $g_i \in H_i$ for $i = 1, \ldots, n$, and $s \in T$. Then $(w_i\theta)^{g_i} \equiv w_i'\theta x \pmod{H_1 \circ \ldots \circ \hat{H}_i \circ \ldots \circ H_n \circ T}$. If $x \notin Z_i$, then $x \in H_j$ for $j \neq i$ and we can say $(w_i\theta)^{g_i} \equiv w_i'\theta \pmod{H_1 \circ \ldots \circ \hat{H}_i \circ \ldots \circ H_n \circ T} \Rightarrow w_i\theta \sim_{H_i} w_i'\theta \pmod{H_i \cap T}$ by the second isomorphism theorem ($AB/A \cong B/(A \cap B)$) with $A := H_1 \circ \ldots \circ \hat{H}_i \circ \ldots \circ H_n \circ T$ and $B := H_i \langle \theta \rangle$ and we are done. Similarly, if $x \in Z_i$, then $(w_i\theta)^{g_i} \equiv w_i'\theta x \pmod{H_1 \circ \ldots \circ \hat{H}_i \circ \ldots \circ H_n \circ T} \Rightarrow w_i\theta \sim_{H_i} w_i'\theta x \pmod{H_i \cap T}$. But $x \in Z_i = H_i \cap T$ so $w_i\theta \sim_{H_i} w_i'\theta \pmod{H_i \cap T}$.

Hence there are $c_1...c_n$ C-classes of elements of $C\theta$ which induce outer automorphisms of C of order two. ∎

Lemma 1.38 Suppose A is a finite perfect group, and E is a central extension of A. Let $\pi : E \to A$ be the natural projection. If $\ker\pi \cap E' = 1$, then $E \cong \ker\pi \times A$.

Proof. Suppose $\ker\pi \cap E' = 1$. Then $\pi|_{E'}$ is an isomorphism, and by Lemma 1.11, $\pi(E') = A$. There is therefore a splitting $s : A \to E$, namely $(\pi|_{E'})^{-1}$, and $E \cong \ker\pi : A$. Since E is a central extension, the product is direct and $E \cong \ker\pi \times A$. ∎

Chapter 2
The Dihedral group of order 6

Lemma 2.1 Suppose y is an element of $\mathcal{T}$ of order 3, and $C := C(y)$. Let r be the number of factors of C, and X_i, $i = 1,...,r$ be the factors of C. If $w \in C$, then $w\theta$ is an involution $\mathrm{mod} Z(C)$ $\Leftrightarrow$ $x_i\theta$ is an involution mod $Z(X_i)$ for $i = 1, ..., r$ where
$$w = \prod_{i=1}^{r} x_i, \text{ and } x_i \in X_i.$$

Proof. ($\Leftarrow$) We proceed by induction on r. Suppose $x_i\theta$ is an involution mod X_i for each $i = 1, ..., r$. Then $x_1 ... x_r \theta x_1 ... x_r \theta =$
$x_1 ... x_r x_1{}^\theta ... x_{r-1}{}^\theta \theta x_r \theta = x_1 ... x_{r-1} x_1{}^\theta ... x_{r-1} x_1 \theta ... x_{r-1}{}^\theta x_r \theta x_r \theta$ (θ fixes all the factors of C (see Table 4 in [CoGr '87]) so x_r commutes with $x_i{}^\theta$ $\forall i \ne r$) =
$x_1 ... x_{r-1} \theta x_1 ... x_{r-1} \theta z_r$ (where $z_r \in Z(X_r)$) $\equiv 1$ mod $Z(C)$ (by induction).

($\Rightarrow$) Suppose conversely that $w\theta$ is an involution mod $Z(C)$: By Table 4 of [CoGr '87], there are three cases:

<u>Case 1</u> (3A): There is nothing to prove here since C has only one factor.

<u>Case 2</u> ($r = 1$, and $C = XT$ where X is the factor and T is a torus contained in $\mathcal{T}$): Since $w \in C$, $w = xt$ where $x \in X$, and $t \in T$. Now $1 = xt\theta xt\theta = x\theta t^{-1}xt\theta = x\theta x\theta$, and $t\theta$ is an involution by Corollary 1.31. Case 1 covers 3C and 3D.

<u>Case 3</u> ($r = 2$, and $C = X_1X_2$ where the X_i are the factors): Then $w = x_1 x_2$ where $x_i \in X_i$. Now $1 = x_1 x_2 \theta x_1 x_2 \theta = x_1 \theta x_2{}^\theta x_1 \theta x_2 \theta = x_1 \theta x_1 x_2 {}^{\theta\theta} x_2 \theta$ (since in this case θ fixes each factor) $= x_1 \theta x_1 \theta x_2 \theta x_2 \theta$. So $(x_1\theta)^2 = (x_2\theta)^{-2}$. Hence $(x_1\theta)^2, (x_2\theta)^2 \in X_1 \cap X_2$, that is, $(x_1 \theta)^2 \equiv (x_2 \theta)^2 \equiv 1$ mod $Z(C)$. This covers 3B. ∎

Theorem 2.2 There are 7 conjugacy classes of subgroups of G which are isomorphic to Dih_6.

Proof. Suppose y is an element of G of order 3, x an involution that inverts y. Then $\langle x, y \rangle \cong \text{Dih}_6$, and $x \in N \backslash C$ where $N := N_G(\langle y \rangle)$, and $C := C_G(y)$. Suppose y' is another element of G of order 3 and x' an involution that inverts y'. If y is not conjugate to y', then $\langle x, y \rangle$ is not conjugate to $\langle x', y' \rangle$. Suppose on the other hand that $y'^g = y$ where $g \in G$. Then $\langle y', x' \rangle^g = \langle y, x'' \rangle$ for some involution x'' inverting y. Since all involutions in Dih_6 are conjugate, $\langle y', x' \rangle$ is conjugate to $\langle y, x \rangle$ if and only if x'' is conjugate in N to x. So the number of conjugacy classes of subgroups of G which are isomorphic to Dih_6 and in which a given conjugacy class of elements of order 3 is represented is the number of conjugacy classes of involutions in $N \backslash C$. But $N/C \cong Z_2$, hence all involutions of $N \backslash C$ are in Cx where x is an involution of $N \backslash C$. So we need to find the number of N-classes of involutions of Cx. We analyze each case separately in the following Lemmas.

Lemma 2.3 If $y \in 3A$, then there is one N-class of involutions in $N \backslash C$, the involutions in this N-class have trace -8 and the connected centralizer of a Dih_6 group with elements in 3A has type B_4.

Proof. Without loss of generality, let $y \in 3A \cap T$. Then C has type A_8. Since $y \in T$, θ inverts y so $\theta \in N \backslash C$. By [Gr '91](2.23), since $-1 \notin W(A_8)$, (the Weyl group of A_8) $\theta |_C$ induces an outer automorphism on C and by [Gr '91](2.18), there is a unique conjugacy class of outer automorphisms of C of order two. Since elements of $C\theta$ induce automorphisms of C and since any outer automorphism of C of order two is represented in $C\theta$, there is a unique C-class of elements of $C\theta$ which induce outer automorphisms of order two, and therefore by Lemma 1.34, there is a unique N-class of elements of $C\theta$ which induce outer automorphisms of order two. Now if $S := \langle y, c\theta \rangle$, then by [Gr '91](2.18), $C_G(S) = C_C(c\theta)$ has type B_4 so its dimension is 36 and must equal $(1, \chi \mid S) = (1/6)(248 + 2(-4) + 3\chi(c\theta))$. Hence $\chi(c\theta) = -8$ and the elements of $C\theta$ which induce outer automorphisms of order two are involutions by Table 1.16, so there is a unique N-class of involutions in $C\theta$. ■

Lemma 2.4 If $y \in 3B$, then there are two N-classes of involutions in $N\backslash C$, and the connected centralizer of a Dih_6 group with elements in 3B has type B_1F_4 in case the involutions have trace 24, and has type B_1C_4 in case the involutions have trace -8.

Proof. Suppose $y \in 3B \cap T$. Then $C = X_1X_2$ where X_1 has type A_2, and X_2 has type E_6. By Lemma 1.32, X_1, X_2 are simply connected. By Corollary 1.5, θ stabilizes each factor of C and by Lemma 1.30, since $y \in T$, θ inverts y. Since $-1 \notin W(A_2)$, $W(E_6)$, $\theta \mid_{X_i}$ is an outer automorphism of X_i for $i = 1, 2$, and by [Gr '91](2.18) there is a unique conjugacy class of outer automorphisms of X_1 of order two, and there are two conjugacy classes of outer automorphisms of X_2 of order two. Since elements of $X_i\theta$ for $i \in \{1, 2\}$, induce automorphisms of X_i, and since any outer automorphism of X_i of order two is represented in $X_i\theta$, there is an unique X_1-class of elements of $X_1\theta$ which induces outer automorphisms of X_1 of order two and there are two X_2-classes of elements of $X_2\theta$ which induce outer automorphisms of X_2 of order two. We want to show: If $x_1, x_1' \in X_1$, and $x_2, x_2' \in X_2$ then $x_1 x_2 \theta \sim_C x_1' x_2' \theta \Leftrightarrow x_1 \theta \sim (X_1/Z(X_1)) x_1' \theta$ and $x_2 \theta \sim (X_2/Z(X_2)) x_2' \theta$.

($\Leftarrow$) Suppose $(x_1 \theta)^{g_1} \equiv x_1' \theta \mod Z(X_1)$ and $(x_2 \theta)^{g_2} \equiv x_2' \theta \mod Z(X_2)$ where $g_1 \in X_1$ and $g_2 \in X_2$. Then $(x_1 x_2 \theta)^{g_2 g_1} \equiv (x_1)^{g_2 g_1}(x_2 \theta)^{g_2 g_1} \mod Z(X_1X_2) \equiv (x_1)^{g_1}(x_2'\theta)^{g_1} \mod Z(X_1X_2) \equiv (x_2')^{g_1}(x_1 \theta)^{g_1} \mod Z(X_1X_2) \equiv x_2' x_1' \theta \mod Z(X_1X_2) \equiv x_1' x_2' \theta \mod Z(X_1X_2)$. Since $y \in T$, θ inverts y and therefore $y^{-1}(y\theta)y = \theta y = y^{-1}\theta$, and $y^{-1}(y^{-1}\theta)y = yy^{-1}\theta = \theta$. Hence, there is only one $\langle y \rangle$-class in $\langle y \rangle \theta$, and since $Z(X_1) = Z(X_2) = \langle y \rangle$, and $C = X_1X_2$,

$x_1 x_2 \theta \sim_C x_1 \,'x_2\, '\theta$.

($\Rightarrow$) Suppose that $(x_1 x_2 \theta)^g = x_1\, 'x_2\, '\theta$, where $g \in C$. Then $g = g_1 g_2$ where $g_1 \in X_1$, and $g_2 \in X_2$. Hence, $(x_2 \theta)^{g_2} \equiv x_2\, '\theta \pmod{X_1}$ and $(x_1 \theta)^{g_1} \equiv x_1\,'\theta \pmod{X_2}$, so $x_1 \theta \sim_{C/X_2} x_1\, '\theta$, and $x_2 \theta \sim_{C/X_1} x_2\, '\theta$. But $C/X_2 \cong X_1/Z(X_1)$ and $C/X_1 \cong X_2/Z(X_2)$ and the result is proved.

Now by Lemma 2.1, there are two C-classes of elements of $C\theta$ which induce outer automorphisms of C of order two. But then by Lemma 1.34, there are two N-classes of elements of $C\theta$ which induce outer automorphisms of C of order two. By [Gr '91](2.18), if $S = \langle y, c\theta \rangle$, $C_G(S) = C_C(c\theta)$ has type $B_1 F_4$ or $B_1 C_4$ and so its dimension is 55 and 39 in the respective cases, and must equal

$$(1, \chi |_S) = (1/6)[248 + 2(5) + 3\chi(c\theta)].$$

Hence θ has trace 24 and -8 in the respective cases, that is, both classes of elements are classes of involutions. Since there are two cases in Table 2.7, there are exactly two N-classes of involutions in $C\theta$. ∎

Lemma 2.5 If $y \in 3C$, then there are two N-classes of involutions in $C\theta$, and the connected centralizer of a Dih_6 group with elements in 3C has type $B_1 B_5$ in case the involutions have trace 24 and has type $B_3 B_3$ in case the involutions have trace -8.

Proof. Without loss of generality, assume $y \in 3C \cap T$. Then $C = XT$, where X has type D_7, and T is a one-dimensional torus. X is simply connected by Lemma 1.32. By Lemma 1.6, θ stabilizes each factor of C, and by Lemma 1.30, θ inverts y. Since $-1 \notin W(D_7)$, $\theta|_X$ is an outer automorphism of X by [Gr '91] (2.23).

<u>Claim</u>: $X \cap T = Z := Z(X)$.

<u>Proof of Claim</u>: Suppose $Z(X) \not\leq T$. Then $Z(X)T$ contains a fours-group. (If $Z(X) = <g>$, and $<g> \cap T = 1$ this is clear. If $<g> \cap T > 1$, say $<g> \cap T = <g_1>$ where $|g_1| = 2$, then $g_0 = gh^{-1}$, (where h is an element of T of order 4 and $h^2 = g_1$), is an involution outside T). But this says that X contains the centralizer of a fours-group in E_8. But by [CoGr '87](3.7) this is impossible.

By [Gr '91] (2.18) there are four conjugacy classes of outer automorphisms of $X/Z(X)$ of order two. Hence by Lemma 1.37, there are four C-classes of elements of $C\theta$ which induce outer automorphisms of C of order two. So by Lemma 1.34, there are therefore four N-classes of elements of $C\theta$ which induce outer automorphisms of C of order two. By [Gr '91] (2.18) if $S = <y, c\theta>$ where $c \in C$, $C_G(S) = C_C(c\theta)$ has type B_6, B_1B_5, B_2B_4, or B_3B_3 and so its dimension is 78, 58, 46, and 42 in the respective cases, and must equal

$(1, \chi|_S) = (1/10)[248 + 2(28 + 50\tau) + 2(28 + 50\ \tau^*) + 5\chi(c\theta)]$.

Hence $c\theta$ has trace 64, 24, 0, and -8 in the respective cases. In case $c\theta$ has trace 64 or 0, by Table 1.16, $c\theta$ is an element of order four, and if $c\theta$ has trace 24 or -8 then $c\theta$ is an involution. So there are two N-classes of involutions in $C\theta$. ∎

Lemma 2.6 If $y \in 3D$, then there are two N-classes of involutions in $C\theta$, and the connected centralizer of a Dih_6 group with elements in 3D has type E_6T_1 if the involutions have trace 24 and has type A_7 if the involutions have trace -8.

Proof. Without loss of generality assume $y \in 3D \cap T$. Then $C = XT$ where X has type $2E_7$ and T is a one-dimensional torus. Now θ induces an outer automorphism on C, but $\theta|_X$ is inner since X doesn't have any outer automorphisms, say $\theta|_X$ induces the inner automorphism corresponding to conjugation by $u \in X$. Then θ inverts T by Lemma 1.30. Now $u^{-1}\theta$ centralizes X so is in $C(X) \cong SL_2(\mathbb{C})$ and we may represent T by $\{\begin{bmatrix} \lambda^{-1} & 0 \\ 0 & \lambda \end{bmatrix} \mid \lambda \in \mathbb{C}\}$. Since u centralizes T, $u^{-1}\theta \in N(T)$ and inverts T. So $u^{-1}\theta = v$ for some $v \leftrightarrow \begin{bmatrix} 0 & \mu \\ -\mu^{-1} & 0 \end{bmatrix}$. We may assume $\mu = 1$ by replacing u^{-1} by $\begin{bmatrix} \mu^{-1} & 0 \\ 0 & \mu \end{bmatrix} u^{-1}$. This induces the same inner auto on X and then $\begin{bmatrix} \mu^{-1} & 0 \\ 0 & \mu \end{bmatrix} u^{-1}\theta = \begin{bmatrix} r^{-1} & 0 \\ 0 & r \end{bmatrix}^2 u^{-1}\theta$ (where $r^2 = \mu$)

$= \begin{bmatrix} r^{-1} & 0 \\ 0 & r \end{bmatrix} u^{-1}\theta \begin{bmatrix} r & 0 \\ 0 & r^{-1} \end{bmatrix} = \begin{bmatrix} r^{-1} & 0 \\ 0 & r \end{bmatrix} \begin{bmatrix} 0 & \mu \\ -\mu^{-1} & 0 \end{bmatrix} \begin{bmatrix} r & 0 \\ 0 & r^{-1} \end{bmatrix} = \begin{bmatrix} 0 & 1 \\ -1 & 0 \end{bmatrix}$. So $\theta = uv = vu$ since $v \in C(X)$, $u \in X$, and $1 = \theta^2 = v^2 u^2$, but $v^2 \ne 1$ so $u^2 \ne 1$, but u^2 induces the trivial automorphism on X since $\theta^2 = 1$ so $u^2 \in Z(X)$, and $|u| = 4$. So u has type 4A[S] or 4H[S]. Suppose x is another element which inverts y, say $x = vu'$, where $u' \in X$. Suppose $u \sim_X u'$. Then $\exists\, h \in X \le C \le N$ such that $h^{-1}uh = u'$. But then $h^{-1}\theta h = h^{-1}vuh = vh^{-1}uh$ (v commutes with h since $v \in C(X)$, $h \in X$) $= vu' = x$. So there are at most 2 N-classes of involutions in $C\theta$ which invert y. But Table 2.7 shows there are exactly two such N-classes. Since θ is inner on X and inverts T, if $S := \langle c\theta, y \rangle$ then $C_G(S) = C_C(c\theta)$ which is the centralizer in X of an element of order 4. But then the connected centralizer has dimension 79 and hence type $E_6 T_1$ in case the involutions of S have trace 24, and dimension 63 and hence type A_7 in case the involutions of S have trace -8. ∎

Table 2.7 Conjugacy classes of Dih_6 subgroups of G.

By Theorem 2.2, each fusion pattern represents a unique conjugacy class. Columns 2 and 3 give elements which generate Dih_6 subgroups of $\mathcal{A}$ with the given fusion pattern. The last column gives the type of the connected centralizer of a Dih_6 of the given fusion pattern. Here $\omega = e^{2\pi i/3}$.

Classes of elements	Element of order 3	Element of order 2	Type of $C_G(S)$
3A; 2B	diag($\omega,\omega^{-1},\omega,\omega^{-1},\omega,\omega^{-1},\omega,\omega^{-1},1$)	diag($\begin{bmatrix} 0 & 1 \\ 1 & 0 \end{bmatrix}^4$,1)	B_4
3B; 2A	diag($\omega,\omega^{-1},\omega,\omega^{-1},\omega,\omega^{-1},1^3$)	diag($\begin{bmatrix} 0 & 1 \\ 1 & 0 \end{bmatrix}^3$,-$1^3$)	B_1F_4
3B; 2B	diag($\omega,\omega^{-1},\omega,\omega^{-1},\omega,\omega^{-1},1^3$)	diag($\begin{bmatrix} 0 & 1 \\ 1 & 0 \end{bmatrix}^3$,-1,$1^2$)	B_1C_4
3C; 2A	diag($\omega,\omega^{-1},\omega,\omega^{-1},1^5$)	diag($\begin{bmatrix} 0 & 1 \\ 1 & 0 \end{bmatrix}^2$,$1^5$)	B_1B_5
3C; 2B	diag($\omega,\omega^{-1},\omega,\omega^{-1},1^5$)	diag($\begin{bmatrix} 0 & 1 \\ 1 & 0 \end{bmatrix}^2$,-$1^2$,$1^3$)	B_3B_3
3D; 2A	diag($\omega,\omega^{-1},1^7$)	diag($\begin{bmatrix} 0 & 1 \\ 1 & 0 \end{bmatrix}$,-1,$1^6$)	E_6T_1
3D; 2B	diag($\omega,\omega^{-1},1^7$)	diag($\begin{bmatrix} 0 & 1 \\ 1 & 0 \end{bmatrix}$,-$1^3$,$1^4$)	A_7

Chapter 3
The Dihedral Group of order 10

Theorem 3.1 There are 13 conjugacy classes of subgroups of G which are isomorphic to Dih_{10}.

Proof. Suppose y is an element of G of order 5, and x an involution that inverts y. Then $<x, y> \cong Dih_{10}$, and $x \in N\backslash C$ where $N := N_G(<y>)$ and $C := C_G(y)$. Suppose y' is another element of G of order 5 and x' an involution that inverts y'. If y is not conjugate to either y' or $(y')^2$, then $<x, y>$ is not conjugate to $<x', y'>$. Suppose y is conjugate to y' in G. Then $\exists g \in G$ such that $(y')^g = y$ and then $<y', x'>^g = <y, x''>$ for some involution x'' inverting y. Since all involutions in Dih_{10} are conjugate, $<y', x'>$ is conjugate to $<y, x>$ if and only if x'' is conjugate in N to x. So the number of conjugacy classes of subgroups of G which are isomorphic to Dih_{10} and in which a given conjugacy class of elements of order 5 is represented is the number of conjugacy classes of involutions in $N\backslash C$. But N/C is isomorphic to a subgroup of Z_4, hence all involutions in $N\backslash C$ are in Cx where x is an involution of $N\backslash C$, so we need to find the number of N-classes of involutions in Cx. We need a few Lemmas:

Lemma 3.2 Suppose y is an element of $\mathcal{T}$ of order 5. Let r be the number of factors of C, and X_i, $i = 1, ..., r$ be the factors of C. If $w \in C$, then $w\theta$ is an involution mod $Z(C) \Leftrightarrow x_i\theta$ is an involution mod $Z(X_i)$ for $i = 1, ..., r$ where $w = \prod_{i=1}^{r} x_i$, and $x_i \in X_i$.

Proof. ($\Leftarrow$) We proceed by induction on r. Suppose $x_i\theta$ is an involution mod X_i for each $i = 1, ..., r$. Then $x_1 ... x_r\theta x_1 ... x_r\theta = x_1 ... x_rx_1 \theta ... x_{r-1}\theta\theta x_r\theta = x_1 ... x_{r-1}x_1 \theta ... x_{r-1}\theta x_r\theta x_r\theta$ (θ fixes all the factors of C (see Table 4 in

[CoGr '87]) so x_r commutes with $x_i\theta$ $\forall i \neq r$) = $x_1 \ldots x_{r-1}\theta x_1 \ldots x_{r-1}\theta z_r$ (where $z_r \varepsilon$ $Z(X_r)$) $\equiv 1$ mod $Z(C)$ by induction.

($\Rightarrow$) Suppose conversely that $w\theta$ is an involution mod $Z(C)$: By Table 4 of [CoGr '87], there are three cases:

<u>Case 1</u> ($r = 1$, and $C = XT$ where X is the factor and T is a torus contained in $\mathcal{T}$): Since $w \varepsilon C$, $w = xt$ where $x \varepsilon X$, and $t \varepsilon T$. Now $1 \equiv xt\theta xt\theta$ (mod $Z(C)$) = $x\theta t^{-1}xt\theta = x\theta x\theta$, and $t\theta$ is an involution by Corollary 1.31. Case 1 covers 5A, 5F, 5G, 5H.

<u>Case 2</u> ($r = 2$, and $C = X_1X_2T$ where the X_i are the factors and T is a torus contained in $\mathcal{T}$): Then $w = x_1 x_2 t$ where $x_i \varepsilon X_i$ and $t \varepsilon T$. Now $1 \equiv x_1 x_2 t\theta x_1 x_2 t\theta$ (mod $Z(C)$) = $x_1 \theta x_2 {}^\theta x_1 \theta x_2 {}^\theta = x_1 \theta x_1 x_2 {}^\theta \theta x_2 {}^\theta$ (since in each of these cases θ fixes each factor) = $x_1 \theta x_1 \theta x_2 \theta x_2 \theta$. So $(x_1\theta)^2 \equiv (x_2\theta)^{-2}$ mod $Z(C)$. Hence $(x_1\theta)^2$, $(x_2\theta)^2 \varepsilon X_1 \cap X_2$. Now case 2 covers 5B, 5D, 5E. But in each of these cases, $(|Z(X_1)|, |Z(X_2)|) = 1$, so $X_1 \cap X_2 = 1$, so $x_1 \theta$, $x_2 \theta$ are involutions.

The only case that remains is:

<u>Case 3</u> ($y \varepsilon$ 5C): As in the previous case we get $(x_1\theta)^2$, $(x_2\theta)^2 \varepsilon X_1 \cap X_2$, that is, $(x_1\theta)^2 \equiv (x_2\theta)^2 \equiv 1$ mod $Z(C)$. ∎

Lemma 3.3 If $y \varepsilon$ 5A, then there is one N-class of involutions in $C\theta$, the involutions in this N-class have trace -8 and the connected centralizer of a Dih_{10} group with elements in 5A has type D_4.

Proof. Without loss of generality, assume $y \varepsilon$ 5A $\cap \mathcal{T}$. Then $C = XT$ where X has type A_7 and T is a one-dimensional torus. By Lemma 1.35, $X \cap T = Z(X)$, and X is simply connected. By Lemma 1.6, θ stabilizes each factor of C and by Lemma 1.30, since $y \varepsilon \mathcal{T}$, θ inverts y, so $\theta \varepsilon N\backslash C$. Since $-1 \notin W(A_7)$, $\theta|_X$ is an outer automorphism of X by [Gr '91] (2.23). By [Gr '91] (2.18) $(X/Z(X))\theta$ has two $(X/Z(X))$-classes of outer automorphisms of order two. By Lemma

1.37, there are two C-classes of elements of $C\theta$ which induce outer automorphisms of order two on C. Hence, by Lemma 1.34, there are 2 N-classes of such elements in $C\theta$. If $S := \langle y, c\theta \rangle$, then by [Gr '91] (2.18), $C_G(S) = C_C(c\theta)$ has type D_4 or C_4 so its dimension is 28 or 36 in the respective cases and equals

$$(1, \chi \mid_S) = (1/10)(248 + 2(8 + 20\tau) + 2(8 + 20\tau^*) + 5\chi(c\theta))$$

where $\tau = (1+\sqrt{5})/2$ and $\tau^* = (1-\sqrt{5})/2$. Hence $c\theta$ has trace -8 and 8 in the respective cases. But trace 8 corresponds to an element of order four, and we're interested in elements of order two. Hence there is only one N-class of involutions in $C\theta$. ∎

Lemma 3.4 If $y \in 5B$ then there is one N-class of involutions in $C\theta$.

Proof. Without loss of generality, assume $y \in 5B \cap T$. Then $C = X_1 X_2 T$ where X_1 has type A_6, X_2 of type A_1, T a one-dimensional torus. By Lemma 1.32, X_1 and X_2 are simply connected, so by Lemma 1.35, $X_1 \cap T = Z(X_1) \cong Z_7$, and $X_2 \cap T = Z(X_2) \cong Z_2$. Now θ stabilizes each factor of C by Lemma 1.6, and by Lemma 1.30, θ inverts y. Since $-1 \notin W(A_6)$, by [Gr '91] (2.23), $\theta \mid_{X_1}$ is an outer automorphism of X_1. By [Gr '91] (2.18), $(X_1/Z(X_1))\theta$ has one $(X_1/Z(X_1))$-class of elements which induce outer automorphisms of $X_1/Z(X_1)$ of order two.

X_2 is simply connected and has type A_1 so $X_2 \cong SL_2(\mathbb{C}) \cong \langle \begin{bmatrix} 1 & t \\ 0 & 1 \end{bmatrix}, \begin{bmatrix} 1 & 0 \\ t & 1 \end{bmatrix} \mid t \in \mathbb{C} \rangle$. The isomorphism is given by: $\begin{bmatrix} 1 & t \\ 0 & 1 \end{bmatrix} \to x_r(t)$, $\begin{bmatrix} 1 & 0 \\ t & 1 \end{bmatrix} \to x_{-r}(t)$. Now $x_r(t)^\theta = x_{-r}(-t)$ so $\begin{bmatrix} 1 & t \\ 0 & 1 \end{bmatrix}^\theta = \begin{bmatrix} 1 & 0 \\ -t & 1 \end{bmatrix}$, i.e. θ is the inverse transpose map. But since θ is an involution, $\begin{bmatrix} a & b \\ c & d \end{bmatrix} \theta$ is an involution $\Rightarrow \begin{bmatrix} a & b \\ c & d \end{bmatrix} \theta \begin{bmatrix} a & b \\ c & d \end{bmatrix} \theta = 1 \Rightarrow \begin{bmatrix} a & b \\ c & d \end{bmatrix} \begin{bmatrix} a & b \\ c & d \end{bmatrix}^\theta = 1 \Rightarrow \theta$ is the inverse map on $\begin{bmatrix} a & b \\ c & d \end{bmatrix}$. So $\begin{bmatrix} a & b \\ c & d \end{bmatrix} \theta$ is an involution if and only if $\begin{bmatrix} a & b \\ c & d \end{bmatrix}\begin{bmatrix} d & -c \\ -b & a \end{bmatrix} = \begin{bmatrix} ad-b^2 & a(b-c) \\ d(b-c) & ad-c^2 \end{bmatrix} = \begin{bmatrix} 1 & 0 \\ 0 & 1 \end{bmatrix}$.

This condition implies $ad - b^2 = 1$, $a(b-c) = 0$, $d(b-c) = 0$, and $ad - c^2 = 1$.

<u>case b-c ≠ 0</u>: In this case, $a = d = 0$ and $b^2 = c^2 = -1 + ad = -1$. So $b = \pm c$, but then $b \neq c \Rightarrow b = -c = \pm i$. But then $ad - bc = ad + b^2 = 0 + -1 = -1$ which is not possible.

<u>case b-c = 0</u>: In this case, $b = c$ and therefore all involutions in $X_2\theta$ are of the form $\begin{bmatrix} a & b \\ b & d \end{bmatrix}\theta$. Suppose $\begin{bmatrix} e & f \\ f & g \end{bmatrix}\theta$ is another involution of $X_2\theta$. Let

$$q := \begin{bmatrix} 1 & 0 \\ (\sqrt{b^2-da+eg} - \sqrt{b^2-d(a-e)})/e & 1 \end{bmatrix}\begin{bmatrix} 1 & (-b + \sqrt{b^2-d(a-e)})/d \\ 0 & 1 \end{bmatrix}.$$

Then $q\begin{bmatrix} a & b \\ b & d \end{bmatrix}\theta q^{-1} =$ $q\begin{bmatrix} a & b \\ b & d \end{bmatrix}q^t\theta = \begin{bmatrix} e & \sqrt{f^2} \\ \sqrt{f^2} & g \end{bmatrix}\theta$, and $\begin{bmatrix} 1 & -2f/g \\ 0 & 1 \end{bmatrix}\begin{bmatrix} e & f \\ f & g \end{bmatrix}\theta\begin{bmatrix} 1 & -2f/g \\ 0 & 1 \end{bmatrix}^{-1} = \begin{bmatrix} e & -f \\ -f & g \end{bmatrix}\theta$, so as all involutions of $X_2\theta$ of type $\begin{bmatrix} a & b \\ b & d \end{bmatrix}\theta$ are conjugate to $\begin{bmatrix} e & f \\ f & g \end{bmatrix}\theta$, there is only one conjugacy class of involutions in $X_2\theta$. So by Lemma 1.37, there is only one C-class of involutions in $C\theta$ which induce outer automorphisms of order two, and therefore only one N-class. Now if $S := \langle y, c\theta \rangle$, $C_G(S) = C_C(c\theta)$ has type B_3T_1 by [Gr '91] (2.18) and so its dimension is 22 and equals

$(1, \chi |_S) = (1/10)[248 + 4(3) + 5\chi(c\theta)]$.

So $\chi(c\theta) = -8$. Hence the involutions in $C\theta$ have type 2B. ∎

Lemma 3.5 If $y \varepsilon 5C$, then there is one N-class of involutions in N\C.

Proof. Without loss of generality, assume $y \varepsilon 5C$. Then $C = X_1X_2$, where X_i is a group of type A_4 for $i = 1, 2$. By Lemma 1.32, X_i is simply connected for $i = 1, 2$. Let x be an involution of G such that x inverts y. Then x normalizes $\langle y \rangle$, hence x normalizes C. But x inverts y so induces an outer automorphism on C. By Corollary 1.5, x permutes X_1 and X_2.

<u>Claim</u>: x is nontrivial on both factors X_i

Proof: Suppose x is trivial on $X_i/\langle y\rangle$ but not on $\langle y\rangle =: Z$. Then $|Z_x| = 5$, $|x| = 2 \Rightarrow \exists$ a fixed point for x in Z_x. So each coset of Z in X_i contains an element of $C_{X_i}(x)$. So $X_i = ZC_{X_i}(x) = Z : C_{X_i}(x)$ (since $C_Z(x) = 1$) $= Z \times C_{X_i}(x)$ since $Z = Z(X_i)$. But $Z \leq X_i' = X_i$, a contradiction. Hence the claim is proved. Want to show: if $x_i, x_i' \in X_i$, for $i = 1, 2$, then $x_1 x_2 x \sim_C x_1' x_2' x \Leftrightarrow x_i x \sim_{X_i} x_i' x$ for $i = 1, 2$.

We get ($\Leftarrow$) by Lemma 1.33. Conversely, assume $x_1 x_2 x \sim_C x_1' x_2' x$. Say $(x_1 x_2 x)^{g_1 g_2} = x_1' x_2' x$. Then $(x_i x)^{g_i} = x_i' x \pmod{X_j}$, $i = 1, 2; j = 3-i$. So $x_i x \sim_{X_i} x_i' x \pmod{Z}$ (since $X_1 X_2 \langle x\rangle / X_j \cong X_i \langle x\rangle / X_j \cap X_i \langle x\rangle = X_i \langle x\rangle / Z$). But x inverts Z so $C_Z(x) = 1$ so x^Z has length 5, so there is only one Z-class of involutions in Z_x. Hence $x_i x \sim_{X_i} x_i' x$ for $i = 1, 2$, and we have ($\Rightarrow$). By [Gr '91] (2.18), there is only one X_i-class of involutions in $X_i x$ so there is only one C-class of involutions in Cx. By [Gr '91] (2.18), $C_{X_i}(x)$ has type B_2 so $C_G(\langle x, y\rangle)$ has dimension 20 and hence

$$20 = (1, \chi|_{\langle x,y\rangle}) = (1/10)[248 + 4(-2) + 5\chi(x)]$$

so $\chi(x) = -8$, and $x \in 2B$. This case occurs in Table 1. ∎

Lemma 3.6 If $y \in 5D$, then there are two N-classes of involutions in $C\theta$.

Proof. If $y \in 5D \cap \mathcal{T}$, then $C = X_1 X_2 T$ where X_1 has type A_2, X_2 has type D_5 and T is a one-dimensional torus. θ stabilizes each factor of C by Lemma 1.6. By [Gr '91] (2.23), since $-1 \notin W(A_2)$, $W(D_5)$, $\theta|_{X_i}$ is an outer automorphism for X_i, $i = 1, 2$. X_1, X_2 are simply connected by Lemma 1.32.

Claim: (1) $X_1 \cap T = Z_1$ where $Z_1 := Z(X_1) \cong \mathbb{Z}_3$.

(2) $X_2 \cap T = Z_2$ where $Z_2 := Z(X_2) \cong \mathbb{Z}_4$.

Proof: (1) Let $r \in Z_1$, $r \neq 1$. Then $r \in X_1 \Rightarrow X_2 T$ centralizes r, and $r \in Z_1$ $\Rightarrow C_G(r)$ has a factor centralizing $X_2 T$ and containing an A_2-group. Hence $r \in$ 3B, $C_G(r) = X_1 V$ where V has type E_6, and $X_2 T$ $C_V(y)$. So, since $X_2 T$ contains a maximal torus of V, $Z(V) \leq X_2T$. Since $|Z(X_2)| = 4$ and $r \in Z(V)$ we have $r \in T$. So $Z_1 \leq T$ and we have (1).

(2) Suppose $X_2 \cap T$ is proper in Z_2. Then X_2T contains a group E of type $Z_4 \times Z_2$. The elements of E of order 4 have centralizer containing $X_1 X_2 T$, so have type 4C, and then the involutions have type 2B. But then $(\chi, 1) = 29$, that is the dimension of $C(E)$ is 29 and $C(E)$ contains C which is a contradiction. So we have (2).

By [Gr '91] (2.18) there is one $(X_1/Z(X_1))$-class of outer automorphisms of $X_1/Z(X_1)$ in $(X_1/Z(X_1))\theta$, and there are three $(X_2/Z(X_2))$-classes of outer automorphisms of $X_2/Z(X_2)$ in $(X_2/Z(X_2))\theta$. Hence by Lemma 1.37 there are three C-classes of elements of $C\theta$ which induce outer automorphisms of C of order 2 and therefore three N-classes of such elements by Lemma 1.34. By [Gr '91] (2.18) if $S = \langle c\theta, y\rangle$, $C_G(S) = C_C(c\theta)$ has type $B_1 \circ B_4$, $B_1 \circ (B_3 \times B_1)$, or $B_1 \circ (B_2 \times B_2)$ and its dimension is 39, 27 and 23 in the respective cases and equals

$(1, \chi|_S) = (1/10)[248 + 2(3 + 5\tau) + 2(3 + 5\tau^*) + 5\chi(c\theta)]$ where $\tau = (1 + \sqrt{5})/2$, $\tau^* = (1 - \sqrt{5})/2$. So $\chi(c\theta) = 24, 0$, and -8 in the respective cases, so the case $C_G(S)$ of type $B_1 \circ (B_3 \times B_1)$ corresponds to an element of order 4. The other cases occur in Table 3.11. Hence there are two N-classes of involutions in $C\theta$. ∎

Lemma 3.7 If $y \in 5E$, then there are two N-classes of involutions in $C\theta$.

Proof. Suppose $y \in 5E \cap T$. Then $C = X_1X_2T$ where X_1 has type A_1, X_2 has type E_6, and T is a one-dimensional torus. θ stabilizes each factor of C by Lemma 1.6. By Lemma 1.30, θ inverts y. By [Gr '91] (2.23), since $-1 \notin W(A_1)$, $W(E_6)$, $\theta|_{X_i}$ is an outer automorphism of X_i for $i = 1, 2$. X_i, $i = 1, 2$ is simply connected by Lemma 1.32.

<u>Claim</u>: (1) $X_1 \cap T = Z_1 := Z(X_1) \cong Z_2$.

(2) $X_2 \cap T = Z_2 := Z(X_2) \cong Z_3$.

Proof. (1) Let $r \in Z_1$, $r \neq 1$. Then X_2T centralizes r and so $C_G(r)$ has a factor centralizing X_2T and containing an A_1-group. Hence $r \in 2A$, and $X_2T \leq Z_{2E_7}(y)$. So $Z(2E_7) \leq X_2T$. $|Z(X_2)| = 3$, $r \in Z(2E_7) \Rightarrow r \in T$, so $Z_1 \leq T$.

(2) Let $r \in Z_2$, $r \neq 1$. $r \in X_2 \Rightarrow X_1T$ centralizes r and so $C_G(r)$ has a factor centralizing X_2T and containing an E_6-group. So $r \in 3B$, and $X_1T \leq 3A_2$. But then $X_1T = C_{3A_2}(y)$ so $Z(3A_2) \leq X_1T$. Since $|Z(X_1)| = 2$ and $r \in Z(3A_2)$, we have $r \in T$ and hence $Z_2 \leq T$.

By [Gr '91] (2.18) there are two $(X_2/Z(X_2))$-classes of outer automorphisms of $X_2/Z(X_2))$ in $(X_2/Z(X_2))\theta$, and one $(X_1/Z(X_1))$-class of outer automorphisms of $X_1/Z(X_1)$ in $(X_1/Z(X_1))\theta$. Hence by Lemma 1.37, there are two C-classes of elements of $C\theta$ which induce outer automorphisms of C of order 2. By Lemma 1.34, there are therefore 2 N-classes of such elements in $C\theta$. Now if $S = \langle \theta, y \rangle$, $C_G(S) = C_C(\theta)$ has type F_4T_1 or C_4T_1 by [Gr '91] (2.18) and so its dimension is 53 or 37 in the respective cases and equals

$(1, \chi |_S) = (1/10)[248 + 2(28 + 25\tau) + 2(28 + 25\tau^*) + 5\chi(\theta)]$ where $\tau = (1 + \sqrt{5})/2$, $\tau^* = (1 - \sqrt{5})/2$. So $\chi(\theta) = 24, -8$ in the respective cases. Both of these cases occur in Table 3.11. ∎

Lemma 3.8 If $y \in 5F$, then there are two N-classes of involutions in $C\theta$.

Proof. Suppose $y \in 5F \cap T$. Then $C = XT$, where X has type D_7, and T is a one-dimensional torus. By Lemma 1.6, θ stabilizes each factor of C. By Lemma 1.30, θ inverts y. By [Gr '91] (2.23), since $-1 \notin W(D_7)$, $\theta|_X$ is an outer automorphism of X. X is simply connected by Lemma 1.32.

<u>Claim:</u> $X \cap T = Z := Z(X)$.

<u>Proof of Claim:</u> Suppose $Z(X)$ is not in T. Then $Z(X)T$ contains a fours-group. (If $Z(X) = \langle g \rangle$, and $\langle g \rangle \cap T = 1$ this is clear. If $\langle g \rangle \cap T > 1$, say $\langle g \rangle \cap T$

$= \langle g_1 \rangle$ where $|g_1| = 2$, then $g_0 = gh^{-1}$ (where h is an element of T of order 4 and $h^2 = g_1$) is an involution outside T). But this says that X contains the centralizer of a fours-group in E_8. But by [CoGr '87](3.7) this is impossible.

By [Gr '91] (2.18) there are four $(X/Z(X))$-classes of outer automorphisms of $X/Z(X)$ of order 2 in $(X/Z(X))\theta$. Hence by Lemma 1.37, there are four C-classes of elements of $C\theta$ which induce outer automorphisms of C of order 2. By Lemma 1.34, there are therefore four N-classes of such elements in $C\theta$. By [Gr '91] (2.18) if $S = \langle y, \theta \rangle$, $C_G(S) = C_C(\theta)$ has type B_6, B_1B_5, B_2B_4, or B_3B_3 and so its dimension is 78, 58, 46, and 42 in the respective cases, and equals

$$(1, \chi|_S) = (1/10)[248 + 2(28 + 50\tau) + 2(28 + 50\ \tau^*) + 5\chi(\theta)].$$

Hence θ has trace 64, 24, 0, and -8 in the respective cases. But the cases where θ has trace 64 and 0 correspond to elements of order 4. The other two cases occur in Table 3.11. Hence there are two N-classes of involutions in $C\theta$. ∎

Lemma 3.9 If $y \in 5G$, then there are two N-classes of involutions in $C\theta$.

Proof. Let $y \in T \cap 5G$. Then $C = XT$ where X has type D_6 and T is a two-dimensional torus. X is simply connected by Lemma 1.32. Since $-1 \in W(D_6)$, $\theta|_X$ is inner, say θ induces conjugation by $u \in X$. Then $u^{-1}\theta$ centralizes X, and therefore centralizes $Z(X) \cong Z_2 \times Z_2$. By [CoGr '87], (Lemma 3.7) since $C_G(Z(X))$ contains a subgroup of type D_6, $Z(X)$ has type BAA and $C_G(Z(X))$ has type $A_1A_1D_6$. $C(X)$ is then of type A_1A_1. $u^{-1}\theta \in C(X)$, and $T \leq C(X)$. Now by [CoGr '87], $C(X) \cong SL_2 \times SL_2$, so $T \leftrightarrow \left\{ \begin{bmatrix} \lambda^{-1} & 0 \\ 0 & \lambda \end{bmatrix} \times \begin{bmatrix} \mu^{-1} & 0 \\ 0 & \mu \end{bmatrix} \right\}$ and

$u^{-1}\theta =: v \leftrightarrow \begin{bmatrix} 0 & \alpha \\ -\alpha^{-1} & 0 \end{bmatrix} \times \begin{bmatrix} 0 & \beta \\ -\beta^{-1} & 0 \end{bmatrix}$ for some $\alpha, \beta \in \mathbb{C}^\times$, since v inverts T (v inverts T since $u^{-1} \in X$ and θ inverts T by Lemma 1.30). Since $u \in X$, u commutes with v so $\theta = uv = vu$. Hence $1 = \theta^2 = u^2v^2$. But $v^2 \neq 1$ so u is not an involution. However $u^2 \in Z(X)$ so is an involution, and v^2 is an involution and $u^2 = v^2$.

But v^2 is a product of two involutions of the form $\begin{bmatrix} -1 & 0 \\ 0 & -1 \end{bmatrix}$ in T, and so is a product of involutions of type 2A in Z(X). Hence $v^2 = u^2$ has type 2B, and therefore u^2 is in the kernel of the map from X onto SO(12, $\mathbb{C}$). Now suppose x' is another involution of G which inverts y. Then since x' is of the form $c\theta$ for $c \in C$, by the argument above, $x' = u'v'$ where $u' \in X$, and $v' \in C(X)$ inverting T.

Claim: $v'u' = x' \sim_C \theta = vu \Leftrightarrow u \sim_X u'$ (mod Z(X))

Proof: Since v, v' are of the form $\begin{bmatrix} 0 & \alpha \\ -\alpha^{-1} & 0 \end{bmatrix} \times \begin{bmatrix} 0 & \beta \\ -\beta^{-1} & 0 \end{bmatrix}$, and since for r, s such that $r^2 = \alpha$ and $s^2 = \beta$ we have $(\begin{bmatrix} r^{-1} & 0 \\ 0 & r \end{bmatrix} \times \begin{bmatrix} s^{-1} & 0 \\ 0 & s \end{bmatrix})(\begin{bmatrix} 0 & \alpha \\ -\alpha^{-1} & 0 \end{bmatrix} \times \begin{bmatrix} 0 & \beta \\ -\beta^{-1} & 0 \end{bmatrix})$ $(\begin{bmatrix} r & 0 \\ 0 & r^{-1} \end{bmatrix} \times \begin{bmatrix} s & 0 \\ 0 & s^{-1} \end{bmatrix}) = \begin{bmatrix} 0 & 1 \\ -1 & 0 \end{bmatrix} \times \begin{bmatrix} 0 & 1 \\ -1 & 0 \end{bmatrix}$, v and v' are conjugate in T. Since T centralizes X, T centralizes u and u' and so we have $v'u' \sim_C vu \Leftrightarrow vu' \sim_C vu$.

($\Rightarrow$) Let $x = vu'$ and suppose $\exists c \in C$ such that $c^{-1}\theta c = x$. Now $c = zt$ where $z \in X$ and $t \in T$, and we have $t^{-1}z^{-1}vuzt = vu'$. But v commutes with z and inverts t, so we get $vtz^{-1}uzt = vu' \Rightarrow t^2z^{-1}uz = u'$ (since z, u, $\in X$, t $\in$ T). Since u', $z^{-1}uz \in X$, $t^2 \in X \cap T = Z(X)$. So $u \sim_X u'$ (mod Z(X)).

($\Leftarrow$) Suppose conversely that $u \sim_X u'$ (mod Z(X)), say $a^{-1}ua = u's$ where a $\in X$ and $s \in Z(X)$. Then $va^{-1}ua = vu's$. Now $v \in C(X)$, so $a^{-1}vua = vu's$. But now let $t \in T$ such that $t^2 = s$. Then $(at)^{-1}vu(at) = t^{-1}(a^{-1}vua)t = t^{-1}(vu's)t = vu'st^2 = vu's^2 = vu'$, and at $\in$ C. So $vu = \theta \sim_C x = vu'$, and the claim is proved.

So, by the claim, we need to find the number of (X/Z(X))-classes of elements of order four in X with square of type 2B in Z(X). Suppose u, u' are elements of X of order four such that $u^2, u'^2 \in Z(X) \cap 2B$. Let $\lambda : X \to SO(12, \mathbb{C})$ be the natural projection. Then $\lambda(u^2) = \lambda(u'^2) = I_{12}$, so u and u' correspond to "odd" involutions of SO(12, $\mathbb{C}$), that is, involutions with spectra $(1^a, -1^b)$ where a, b $\equiv 2$ (mod 4) (see [Gr '91](2.8)). There are three SO(12, $\mathbb{C}$)-classes of such involutions, namely, those with spectra $(1^2, -1^{10})$, $(1^6, -1^6)$ and $(1^{10}, -1^2)$. If $\lambda(u)$ is conjugate to $\lambda(u')$, say by g, then u is conjugate to u' mod Z(X) in X so we have at most three X-classes of such elements. But note that involutions with spectra $(1^2, -1^{10})$ and $(1^{10}, -1^2)$ differ by a factor of $-I_{12}$ which lifts to an element of Z(X), so there are in fact two X-classes of elements of order four in X which square to $Z(X) \cap 2B$, and therefore two G-classes of Dih_{10}-subgroups of G with elements of type 5G. ∎

Lemma 3.10 If y ε 5H, then there are two N-classes of involutions in Cθ.

Proof. Suppose y ε $T \cap$ 5H. Then C = XT where X has type E_7 and T is a one-dimensional torus. Then θ induces an outer automorphism on C. $\theta|_X$ is inner since E_7 doesn't have any outer automorphisms, so say $\theta|_X$ is conjugation by u ε X. Hence $\theta|_T$ is an outer automorphism of T, so θ inverts T. Now $u^{-1}\theta$ centralizes X so is in $2A_1 \cong SL_2(\mathbb{C})$ [since $(2A_1)(2E_7) = $ C(involution in $Z(2E_7)) \supseteq C(2E_7)$ and $u^{-1}\theta \notin 2E_7$] and we may represent T by $\{\begin{bmatrix} \lambda^{-1} & 0 \\ 0 & \lambda \end{bmatrix} \mid \lambda \varepsilon \mathbb{C} \}$. Since u centralizes T, $u^{-1}\theta \varepsilon N(T)$ and inverts T. So $u^{-1}\theta = v$ for some $v \leftrightarrow \begin{bmatrix} 0 & \mu \\ -\mu^{-1} & 0 \end{bmatrix}$. We may assume $\mu = 1$ by replacing u^{-1} by $\begin{bmatrix} \mu^{-1} & 0 \\ 0 & \mu \end{bmatrix} u^{-1}$. This induces the same inner auto on $2E_7$ and then $\begin{bmatrix} \mu^{-1} & 0 \\ 0 & \mu \end{bmatrix} u^{-1}\theta = \begin{bmatrix} r^{-1} & 0 \\ 0 & r \end{bmatrix}^2 u^{-1}\theta$ (where $r^2 = \mu$) $= \begin{bmatrix} r^{-1} & 0 \\ 0 & r \end{bmatrix} u^{-1}\theta \begin{bmatrix} r & 0 \\ 0 & r^{-1} \end{bmatrix} = \begin{bmatrix} r^{-1} & 0 \\ 0 & r \end{bmatrix} \begin{bmatrix} 0 & \mu \\ -\mu^{-1} & 0 \end{bmatrix} \begin{bmatrix} r & 0 \\ 0 & r^{-1} \end{bmatrix} = \begin{bmatrix} 0 & 1 \\ -1 & 0 \end{bmatrix}$. So x = uv = vu since v ε $2A_1$, u ε $2E_7$, and $1 = \theta^2 = v^2 u^2$, but $v^2 \neq 1$ so $u^2 \neq 1$, but u^2 induces the trivial automorphism on X since $\theta^2 = 1$ so $u^2 \varepsilon Z(X)$, and $|u| = 4$. Now $Z(X) = \{1, h_\beta(-1)\}$ where β is the root (0, 0, 0, 0, 0, 0, 0, 1) as in [CoGr '87]. u is in 4A or 4H of $2E_7$. So if x, x' are two involutions of G which invert y, x = vu, x' = vu', where v is as above, and u, u' are elements of the

2E_7-factor of C and each belongs to one of 4A or 4H in 2E_7. Suppose $u \sim_{2E_7} u'$. Then $\exists$ h ε 2$E_7 \leq$ C $\leq$ N such that $h^{-1}uh = u'$. But then $h^{-1}xh = h^{-1}vuh = vh^{-1}uh$ (v commutes with h since v ε 2A_1, h ε 2E_7) = $vu' = x'$. So there are at most 2 N-classes of involutions in Cx which invert y. But Table 3.11 shows there are exactly two such N-classes. Since θ is inner on X and inverts T, if S := $<c\theta, y>$ then $C_G(S) = C_C(c\theta)$ which is the centralizer in X of an element of order 4. But then the connected centralizer has dimension 79 and hence type E_6T_1 in case the involutions of S have trace 24, and dimension 63 and hence type A_7 in case the involutions of S have trace -8. ∎

Table 3.11 Conjugacy classes of Dih$_{10}$ subgroups of G.

By Theorem 3.1, each fusion pattern represents a unique conjugacy class. Columns 2 and 3 give elements which generate Dih$_{10}$ subgroups of $\mathcal{A}$ with the given fusion pattern. The last column gives the type of the connected centralizer of a Dih$_{10}$ of the given fusion pattern. Here $\rho = e^{2\pi i/5}$.

Classes of elements	Element of order 5	Element of order 2	Type of $C_G(S)$
5A; 2B	diag($\rho, \rho^{-1}, \rho, \rho^{-1}, \rho, \rho^{-1}, \rho, \rho^{-1}, 1$)	diag($\begin{bmatrix} 0 & 1 \\ 1 & 0 \end{bmatrix}^4, 1$)	D_4
5B; 2B	diag($\rho, \rho^{-1}, \rho, \rho^{-1}, \rho, \rho^{-1}, \rho^2, \rho^3, 1$)	diag($\begin{bmatrix} 0 & 1 \\ 1 & 0 \end{bmatrix}^4, 1$)	B_3T_1
5C; 2B	diag($\rho, \rho^{-1}, \rho, \rho^{-1}, \rho^2, \rho^3, \rho^2, \rho^3, 1$)	diag($\begin{bmatrix} 0 & 1 \\ 1 & 0 \end{bmatrix}^4, 1$)	B_2B_2
5D; 2A	diag($\rho, \rho^{-1}, \rho, \rho^{-1}, \rho^2, \rho^3, 1^3$)	diag($\begin{bmatrix} 0 & 1 \\ 1 & 0 \end{bmatrix}^3, -1^3$)	B_1B_4
5D; 2B	diag($\rho, \rho^{-1}, \rho, \rho^{-1}, \rho^2, \rho^3, 1^3$)	diag($\begin{bmatrix} 0 & 1 \\ 1 & 0 \end{bmatrix}^3, -1, 1^2$)	$B_1(B_2 \times B_2)$
5E; 2A	diag($\rho, \rho^{-1}, \rho, \rho^{-1}, \rho, \rho^{-1}, 1^3$)	diag($\begin{bmatrix} 0 & 1 \\ 1 & 0 \end{bmatrix}^3, -1^3$)	F_4T_1
5E; 2B	diag($\rho, \rho^{-1}, \rho, \rho^{-1}, \rho, \rho^{-1}, 1^3$)	diag($\begin{bmatrix} 0 & 1 \\ 1 & 0 \end{bmatrix}^3, -1, 1^2$)	C_4T_1

5F; 2A	$\mathrm{diag}(\rho,\rho^{-1},\rho,\rho^{-1},1^5)$	$\mathrm{diag}(\begin{bmatrix}0 & 1\\1 & 0\end{bmatrix}^2,1^5)$	$B_1 \times B_5$
5F; 2B	$\mathrm{diag}(\rho,\rho^{-1},\rho,\rho^{-1},1^5)$	$\mathrm{diag}(\begin{bmatrix}0 & 1\\1 & 0\end{bmatrix}^2,-1^2,1^3)$	$B_3 \times B_3$
5G; 2A	$\mathrm{diag}(\rho,\rho^{-1},\rho^2,\rho^3,1^5)$	$\mathrm{diag}(\begin{bmatrix}0 & 1\\1 & 0\end{bmatrix}^2,1^5)$	D_5T_1
5G; 2B	$\mathrm{diag}(\rho,\rho^{-1},\rho^2,\rho^3,1^5)$	$\mathrm{diag}(\begin{bmatrix}0 & 1\\1 & 0\end{bmatrix}^2,-1^2,1^3)$	A_3A_3
5H; 2A	$\mathrm{diag}(\rho,\rho^{-1},1^7)$	$\mathrm{diag}(\begin{bmatrix}0 & 1\\1 & 0\end{bmatrix},-1,1^6)$	E_6T_1
5H; 2B	$\mathrm{diag}(\rho,\rho^{-1},1^7)$	$\mathrm{diag}(\begin{bmatrix}0 & 1\\1 & 0\end{bmatrix},-1^3,1^4)$	A_7

Chapter 4
The Alt$_5$ and SL(2, 5) fusion patterns in G, $\mathcal{A}$, Δ and Ω

To consider Alt$_5$ and SL(2, 5)-subgroups of G, we consider the possible fusion patterns of such subgroups which we define below.

Definition 4.1 A *fusion pattern* is a function from the conjugacy classes of one group to the conjugacy classes of another that preserves the power maps. We denote the fusion patterns from Alt$_5$ (resp. SL(2, 5)) to G as a list of the images in G of the classes of Alt$_5$ (resp. SL(2, 5)) in the order in which they appear in [Atlas '85].

Now Alt$_5$ has five conjugacy classes of elements (containing elements of orders 1, 2, 3, 5 and 5 respectively) each of which maps into a single conjugacy class of G. Since there are 2 conjugacy classes of elements of order 2, 4 classes of elements of order 3, and 14 classes of elements of order 5 in G, there are $(2)4(14)^2 = 1568$ quintuples of conjugacy classes $(K_1, K_2, K_3, K_5, L_5)$ where K_i is a conjugacy class of elements of order i in G and L_5 is a conjugacy class of elements of order 5 in G.

We will be considering representations of the groups Alt$_5$, and SL(2,5) in what follows, and it will be useful to know the eigenvalues of the images of elements of these groups. Here we will be using the information on these groups from [Atlas '85]. We begin with Alt$_5$, which has irreducible representations $1_a, 3_a, 3_b, 4_a,$ and 5_a.

Note 4.2 We will at times blur the distinction between a representation, or a class of representations and its character. For example, when we speak of the representation 3_a of Alt_5, we mean either the class of representations which affords the 3_a character of Alt_5, or a representative of that class.

Table 4.3 Elements of Alt_5, their traces and their eigenvalues.

We only give the eigenvalues for one class of elements of order 5 as the eigenvalues for the other class are simply the squares of the eigenvalues of the given class. Here $\omega = e^{2\pi i/3}$ and $\rho = e^{2\pi i/5}$.

Representation	Trace on class of order 2	Eigenvalues	Trace on class of order 3	Eigenvalues	Trace on class of order 5	Eigenvalues
3_a	-1	-1 -1 1	0	$\omega, \omega^{-1}, 1$	$(1-\sqrt{5})/2$	$\rho^2, \rho^3, 1$
3_b	-1	-1 -1 1	0	$\omega, \omega^{-1}, 1$	$(1+\sqrt{5})/2$	$\rho, \rho^{-1}, 1$
4_a	0	-1 -1 1 1	1	$\omega, \omega^{-1}, 1, 1$	-1	$\rho, \rho^{-1}, \rho^2, \rho^3$
5_a	1	-1 -1 1 1 1	-1	$\omega, \omega^{-1}, \omega, \omega^{-1}, 1$	0	$\rho, \rho^{-1}, \rho^2, \rho^3, 1$

Proof. Everything on this Table is from [Atlas '85] except the eigenvalues. For eigenvalues on elements of order 2 the proof is trivial. For elements of order 3, we note that the character of an irreducible representation is always real, so if ω occurs as an eigenvalue n times, so does ω^{-1} and vice versa. Now $\omega + \omega^{-1} = -1$, so it follows easily that the eigenvalues are as in Table 4.3. Finally, for elements of order 5, we note that the character of an irreducible representation is always real. Now the imaginary part of ρ^2 is $\sin(4\pi/5) = 2\cos(2\pi/5)\sin(2\pi/5)$ which is not an integer multiple of $\sin(2\pi/5)$, the imaginary part of ρ. Hence ρ and ρ^{-1} have the same multiplicity, and the same can be said for ρ^2 and ρ^3. Now, since $\rho + \rho^{-1} + \rho^2 + \rho^3 = -1$, it follows easily that the eigenvalues for elements of order five are as in Table 4.3. ∎

Table 4.4 Elements of SL(2, 5), their traces, and their eigenvalues.

Similar comments as in Table 4.3 apply for elements of orders 5 and 10. Here $i = \sqrt{-1}$, $\omega = e^{2\pi i/3}$, $\alpha = e^{2\pi i/6}$, $\rho = e^{2\pi i/5}$, and $\xi = e^{2\pi i/10}$.

Representation	Trace on class of order 4	Eigenvalues	Trace on class of order 3	Eigenvalues	Eigenvalues of elements of order 6	Trace on class of order 5	Eigenvalues	Eigenvalues of elements of order 10
2_a	0	i $-i$	-1	ω ω^{-1}	α α^{-1}	$(-1+\sqrt{5})/2$	ρ ρ^{-1}	ξ^3 ξ^7
2_b	0	i $-i$	-1	ω ω^{-1}	α α^{-1}	$(-1-\sqrt{5})/2$	ρ^2 ρ^3	ξ ξ^{-1}
4_a	0	i i $-i$ $-i$	1	ω ω^{-1} 1 1	α α^{-1} -1 -1	-1	ρ ρ^{-1} ρ^2 ρ^3	ξ ξ^{-1} ξ^3 ξ^7
6_a	0	i i i $-i$ $-i$ $-i$	0	ω ω^{-1} ω ω^{-1} 1 1	α α^{-1} α α^{-1} -1 -1	1	ρ ρ^{-1} ρ^2 ρ^3 1 1	ξ ξ^{-1} ξ^3 ξ^7 -1 -1

Proof. For elements of order 3 and 5, the arguments are the same as in the proof of Table 4.3. For elements of order 6, we note that the square of an element of order 6 is an element of order 3. Hence, for example, the square of an element of order 6 in the image of the representation 2_a has eigenvalues ω and ω^{-1}. So the set of eigenvalues of the element of order 6 consists of one element from $\{\alpha, \omega^{-1}\}$ and one element from $\{\alpha^{-1}, \omega\}$. But the cube of the element has eigenvalues $-1, -1$, so the eigenvalues for the element of order 6 are primitive 6th roots of unity. In fact whenever ω (resp. ω^{-1}) occurs as an eigenvalue for an element of order three, that element's square root then has α (resp. $α^{-1}$) as an eigenvalue. Finally, if 1 occurs as an eigenvalue for an element of order three, that element's square root has -1 as an eigenvalue. Similar arguments lead to the spectra given for elements of orders 4 and 10. ∎

Let L be a subgroup of G such that $L \cong Alt_5$ and consider the adjoint character $χ$ of G. Now $χ|_L$ is a character of L, and therefore the inner product of $χ|_L$ with any irreducible character of L is a nonnegative integer. Hence, we eliminate from consideration any fusion pattern for which $(χ|_L, η)$ is not a nonnegative integer, where $η$ is an irreducible character of L. We also eliminate any fusion pattern in which a nonrational element of order five is in the same class as its square.

Table 4.5 Fusion patterns of $L \cong Alt_5$ in G for which $(χ|_L, η)$ is a nonnegative integer for any irreducible character $η$ of L.

The numbers in the last column represent the inner product of $χ|_L$ with the irreducible characters $1_a, 3_a, 3_b, 4_a$, and 5_a ([Atlas '85] notation) respectively. The information on this Table is due to Robert L. Griess and was checked independently by this author.

Fusion Pattern Number	Classes in G	Inner products with $1, 3_a, 3_b, 4, 5$
1	2A, 3A, 5A, 5A[2]	16, 0, 20, 8, 28
14	2A, 3A, 5A[2], 5A	16, 20, 0, 8, 28
31	2A, 3A, 5B, 5B[2]	10, 7, 7, 14, 28
44	2A, 3A, 5B[2], 5B	10, 7, 7, 14, 28
60	2A, 3A, 5C, 5C	8, 6, 6, 16, 28
76	2A, 3A, 5D, 5D[2]	11, 5, 10, 13, 28
89	2A, 3A, 5D[2], 5D	11, 10, 5, 13, 28
165	2A, 3A, 5G, 5G	18, 11, 11, 6, 28
197	2A, 3B, 5A, 5A[2]	19, 0, 20, 11, 25
210	2A, 3B, 5A[2], 5A	19, 20, 0, 11, 25
227	2A, 3B, 5B, 5B[2]	13, 7, 7, 17, 25
240	2A, 3B, 5B[2], 5B	13, 7, 7, 17, 25

256	2A, 3B, 5C, 5C	11, 6, 6, 19, 25
272	2A, 3B, 5D, 5D[2]	14, 5, 10, 16, 25
285	2A, 3B, 5D[2], 5D	14, 10, 5, 16, 25
302	2A, 3B, 5E, 5E[2]	28, 2, 27, 2, 25
315	2A, 3B, 5E[2], 5E	28, 27, 2, 2, 25
361	2A, 3B, 5G, 5G	21, 11, 11, 9, 25
393	2A, 3C, 5A, 5A[2]	22, 0, 20, 14, 22
406	2A, 3C, 5A[2], 5A	22, 20, 0, 14, 22
423	2A, 3C, 5B, 5B[2]	16, 7, 7, 20, 22
436	2A, 3C, 5B[2], 5B	16, 7, 7, 20, 22
452	2A, 3C, 5C, 5C	14, 6, 6, 22, 22
468	2A, 3C, 5D, 5D[2]	17, 5, 10, 19, 22
481	2A, 3C, 5D[2], 5D	17, 10, 5, 19, 22
498	2A, 3C, 5E, 5E[2]	31, 2, 27, 5, 22
511	2A, 3C, 5E[2], 5E	31, 27, 2, 5, 22
557	2A, 3C, 5G, 5G	24, 11, 11, 12, 22
589	2A, 3D, 5A, 5A[2]	43, 0, 20, 35, 1
602	2A, 3D, 5A[2], 5A	43, 20, 0, 35, 1
619	2A, 3D, 5B, 5B[2]	37, 7, 7, 41, 1
632	2A, 3D, 5B[2], 5B	37, 7, 7, 41, 1
648	2A, 3D, 5C, 5C	35, 6, 6, 43, 1
664	2A, 3D, 5D, 5D[2]	38, 5, 10, 40, 1
677	2A, 3D, 5D[2], 5D	38, 10, 5, 40, 1
694	2A, 3D, 5E, 5E[2]	52, 2, 27, 26, 1
707	2A, 3D, 5E[2], 5E	52, 27, 2, 26, 1
753	2A, 3D, 5G, 5G	45, 11, 11, 33, 1
769	2A, 3D, 5H, 5H[2]	78, 0, 55, 0, 1
782	2A, 3D, 5H[2], 5H	78, 55, 0, 0, 1
785	2B, 3A, 5A, 5A[2]	8, 8, 28, 8, 20
798	2B, 3A, 5A[2], 5A	8, 28, 8, 8, 20
815	2B, 3A, 5B, 5B[2]	2, 15, 15, 14, 20
828	2B, 3A, 5B[2], 5B	2, 15, 15, 14, 20
844	2B, 3A, 5C, 5C	0, 14, 14, 16, 20
860	2B, 3A, 5D, 5D[2]	3, 13, 18, 13, 20
873	2B, 3A, 5D[2], 5D	3, 18, 13, 13, 20
949	2B, 3A, 5G, 5G	10, 19, 19, 6, 20
981	2B, 3B, 5A, 5A[2]	11, 8, 28, 11, 17
994	2B, 3B, 5A[2], 5A	11, 28, 8, 11, 17
1011	2B, 3B, 5B, 5B[2]	5, 15, 15, 17, 17
1024	2B, 3B, 5B[2], 5B	5, 15, 15, 17, 17
1040	2B, 3B, 5C, 5C	3, 14, 14, 19, 17
1056	2B, 3B, 5D, 5D[2]	6, 13, 18, 16, 17
1069	2B, 3B, 5D[2], 5D	6, 18, 13, 16, 17
1086	2B, 3B, 5E, 5E[2]	20, 10, 35, 2, 17
1099	2B, 3B, 5E[2], 5E	20, 35, 10, 2, 17
1145	2B, 3B, 5G, 5G	13, 19, 19, 9, 17

1177	2B, 3C, 5A, 5A[2]	14, 8, 28, 14, 14
1190	2B, 3C, 5A[2], 5A	14, 28, 8, 15, 14
1207	2B, 3C, 5B, 5B[2]	8, 15, 15, 20, 14
1220	2B, 3C, 5B[2], 5B	8, 15, 15, 20, 14
1236	2B, 3C, 5C, 5C	6, 14, 14, 22, 14
1252	2B, 3C, 5D, 5D[2]	9, 13, 18, 19, 14
1265	2B, 3C, 5D[2], 5D	9, 18, 13, 19, 14
1282	2B, 3C, 5E, 5E[2]	23, 10, 35, 5, 14
1295	2B, 3C, 5E[2], 5E	23, 35, 10, 5, 14
1312	2B, 3C, 5F, 5F[2]	28, 0, 50, 0, 14
1325	2B, 3C, 5F[2], 5F	28, 50, 0, 0, 14
1341	2B, 3C, 5G, 5G	16, 19, 19, 12, 14

Lemma 4.6 It is impossible for L to have elements of type 5A, 5B or 5C if the involutions of L have type 2A. Hence we eliminate from consideration fusion patterns: 1, 14, 31, 44, 60, 197, 210, 227, 240, 256, 393, 406, 423, 436, 452, 589, 602, 619, 632 and 648.

Proof. If L has such a fusion pattern, then a maximal Dih_{10} subgroup D of L has a fusion pattern which does not appear on Table 3.11, and is therefore impossible. ∎

Lemma 4.7 It is impossible for L to have elements of type 3A if the involutions of L have type 2A. Hence we eliminate from consideration fusion patterns: 76, 89 and 165.

Proof. If L has such a fusion pattern, then a maximal Dih_6 subgroup D of L has a fusion pattern which does not appear on Table 2.7, and is therefore impossible. ∎

Remark 4.8 Note from Table 4.5 that certain pairs of fusion patterns contain the same classes of elements, but in a different order, e.g. fusion patterns 272 and 285. Suppose $L \cong Alt_5$ and η is an embedding of L into G such that $\eta(L)$ has fusion pattern 272. Let σ be an outer automorphism of L. Then σ

interchanges the two classes of elements of order 5 in L so $\eta(\sigma(L))$ has fusion pattern 285. But $\eta(L) = \eta(\sigma(L))$, so although η and $(\eta \circ \sigma)$ are nonconjugate embeddings of L into G, the images of η and $(\eta \circ \sigma)$ are equal and therefore conjugate. So, when dealing with matters of conjugacy of Alt_5, we treat only one of these pairs. For example, from the class {272, 285}, we treat fusion pattern 272 but not 285. Hence, as we list the remaining fusion patterns, such pairs will be listed together, the twin fusion pattern being listed in parentheses; from now on, we only treat one member of each pair: 272(285), 302(315), 361, 468(481), 498(511), 557, 664(677), 694(707), 753, 769(782), 785(798), 815(828), 844, 860(873), 949, 981(994), 1011(1024), 1040, 1056(1069), 1086(1099), 1145, 1177(1190), 1207(1220), 1236, 1252(1265), 1282(1295), 1312(1325) and 1341. An analogous statement is true for SL(2, 5). The finer problem of classifying nonconjugate embeddings of Alt_5 and SL(2, 5) into G will be addressed in chapters 7 and 8 following Theorems 7.5 and 8.1.

Table 4.9 Fusion patterns of $M \cong SL(2, 5)$ in G for which $(\chi \mid_M, \eta)$ is a nonnegative integer for every irreducible character η of M.

The numbers in the last column represent the inner product of $\chi \mid_M$ with the irreducible characters 1_a, 2_a, 2_b, 4_a(faithful), 6_a, 3_a, 3_b, 4_b(nonfaithful), and 5_a ([Atlas '85] notation) respectively. Each entry in the Table represents two fusion patterns (except for those indicated by a * which represent a single fusion pattern). In light of Remark 4.8, since the two fusion patterns represented by non-starred entries represent two nonconjugate embeddings of the same group, we treat only one of the two fusion patterns.

Fusion pattern number	Classes in G	Inner products with 1_a, 2_a, 2_b, 4_a, 6_a, 3_a, 3_b, 4_b, and 5_a
1	2A, 3A, 4A, 5A, 6C, 10E	16, 7, 27, 7, 7, 0, 27, 1, 7
2	2A, 3A, 4A, 5A, 6C, 10L	11, 12, 17, 2, 12, 5, 17, 6, 7
3	2A, 3A, 4A, 5A, 6C, 10N	11, 16, 13, 2, 12, 9, 13, 6, 7
8	2A, 3A, 4A, 5A, 6G, 10L	7, 8, 17, 6, 12, 5, 17, 2, 11
19	2A, 3A, 4A, 5B, 6C, 10D	17, 28, 0, 14, 0, 28, 0, 0, 7
20	2A, 3A, 4A, 5B, 6C, 10P	11, 18, 4, 8, 6, 18, 4, 6, 7
21	2A, 3A, 4A, 5B, 6G, 10R	10, 13, 8, 7, 7, 13, 8, 7, 7
22	2A, 3A, 4A, 5B, 6C, 10Z	7, 10, 8, 4, 10, 10, 8, 10, 7
23	2A, 3A, 4A, 5B, 6C, 10BB	6, 9, 8, 3, 11, 9, 8, 11, 7
24	2A, 3A, 4A, 5B, 6C, 10JJ	5, 8, 8, 2, 12, 8, 8, 12, 7
26	2A, 3A, 4A, 5B, 6G, 10P	7, 14, 4, 1, 2, 6, 18, 4, 2, 11
27	2A, 3A, 4A, 5B, 6G, 10R	6, 9, 8, 11, 7, 13, 8, 3, 11
28	2A, 3A, 4A, 5B, 6G, 10Z	3, 6, 8, 8, 10, 10, 8, 6, 11
29	2A, 3A, 4A, 5B, 6G, 10BB	2, 5, 8, 7, 11, 9, 8, 7, 11
30	2A, 3A, 4A, 5B, 6G, 10JJ	1, 4, 8, 6, 12, 8, 8, 8, 11
37	2A, 3A, 4A, 5C, 6C, 10U	9, 16, 5, 8, 6, 15, 5, 8, 7
38	2A, 3A, 4A, 5C, 6C, 10KK	5, 10, 7, 4, 10, 9, 7, 12, 7
43	2A, 3A, 4A, 5C, 6G, 10U	5, 12, 5, 12, 6, 15, 5, 4, 11
44	2A, 3A, 4A, 5C, 6G, 10KK	1, 6, 7, 8, 10, 9, 7, 8, 11
55	2A, 3A, 4A, 5D, 6A, 10X	8, 7, 14, 4, 10, 5, 14, 9, 7
56	2A, 3A, 4A, 5D, 6C, 10DD	6, 9, 10, 2, 12, 7, 10, 11, 7
57	2A, 3A, 4A, 5D, 6C, 10HH	6, 10, 9, 2, 12, 8, 9, 11, 7
58	2A, 3A, 4A, 5D, 6C, 10YY	16, 20, 9, 12, 2, 18, 9, 1, 7
59	2A, 3A, 4A, 5D, 6C, 10ZZ	16, 4, 25, 12, 2, 2, 25, 1, 7
60	2A, 3A, 4A, 5D, 6C, 10AAA	8, 12, 9, 4, 10, 10, 9, 9, 7
61	2A, 3A, 4A, 5D, 6G, 10X	4, 3, 14, 8, 10, 5, 14, 5, 11
62	2A, 3A, 4A, 5D, 6G, 10DD	2, 5, 10, 6, 12, 7, 10, 7, 11
63	2A, 3A, 4A, 5D, 6G, 10HH	2, 6, 9, 6, 12, 8, 9, 7, 11
66	2A, 3A, 4A, 5D, 6G, 10AAA	4, 8, 9, 8, 10, 10, 9, 5, 11
112	2A, 3A, 4A, 5G, 6C, 10GGG	15, 13, 9, 4, 10, 17, 9, 2, 7
115	2A, 3A, 4A, 5G, 6G, 10Q	6, 4, 9, 3, 15, 12, 9, 3, 11
116*	2A, 3A, 4A, 5G, 6G, 10TT	5, 2, 10, 2, 16, 10, 10, 4, 11
146	2A, 3A, 4D, 5A, 6C, 10L	13, 12, 15, 2, 12, 3, 15, 6, 9
147	2A, 3A, 4D, 5A, 6C, 10N	13, 16, 11, 2, 12, 7, 11, 6, 9
152	2A, 3A, 4D, 5A, 6G, 10L	9, 8, 15, 6, 12, 3, 15, 2, 13
164	2A, 3A, 4D, 5B, 6C, 10P	13, 18, 2, 8, 6, 16, 2, 6, 9
165	2A, 3A, 4D, 5B, 6C, 10R	12, 13, 6, 7, 7, 11, 6, 7, 9
166	2A, 3A, 4D, 5B, 6C, 10Z	9, 10, 6, 4, 10, 8, 6, 10, 9
167	2A, 3A, 4D, 5B, 6C, 10BB	8, 9, 6, 3, 11, 7, 6, 11, 9
168	2A, 3A, 4D, 5B, 6C, 10JJ	7, 8, 6, 2, 12, 6, 6, 12, 9
170	2A, 3A, 4D, 5B, 6G, 10P	9, 14, 2, 12, 6, 16, 2, 2, 13
171	2A, 3A, 4D, 5B, 6G, 10R	8, 9, 6, 11, 7, 11, 6, 3, 13

CONJUGACY OF Alt$_5$ AND SL(2, 5) SUBGROUPS OF E$_8$($\mathbb{C}$) 53

172	2A, 3A, 4D, 5B, 6G, 10Z	5, 6, 6, 8, 10, 8, 6, 6, 13
173	2A, 3A, 4D, 5B, 6G, 10BB	4, 5, 6, 7, 11, 7, 6, 7, 13
174	2A, 3A, 4D, 5B, 6G, 10JJ	3, 4, 6, 6, 12, 6, 6, 8, 13
181	2A, 3A, 4D, 5C, 6C, 10U	11, 16, 3, 8, 6, 13, 3, 8, 9
182	2A, 3A, 4D, 5C, 6C, 10KK	7, 10, 5, 4, 10, 7, 5, 12, 9
187	2A, 3A, 4D, 5C, 6G, 10U	7, 12, 3, 12, 6, 13, 3, 4, 13
188	2A, 3A, 4D, 5C, 6G, 10KK	3, 6, 5, 8, 10, 7, 5, 8, 13
199	2A, 3A, 4D, 5D, 6C, 10X	10, 7, 12, 4, 10, 3, 12, 9, 9
200	2A, 3A, 4D, 5D, 6C, 10DD	8, 9, 8, 2, 12, 5, 8, 11, 9
201	2A, 3A, 4D, 5D, 6C, 10HH	8, 10, 7, 2, 12, 6, 7, 11, 9
202	2A, 3A, 4D, 5D, 6C, 10YY	18, 20, 7, 12, 2, 16, 7, 1, 9
203	2A, 3A, 4D, 5D, 6C, 10ZZ	18, 4, 23, 12, 2, 0, 23, 1, 9
204	2A, 3A, 4D, 5D, 6C, 10AAA	10, 12, 7, 4, 10, 8, 7, 9, 9
205	2A, 3A, 4D, 5D, 6G, 10X	6, 3, 12, 8, 10, 3, 12, 5, 13
206	2A, 3A, 4D, 5D, 6G, 10DD	4, 5, 8, 6, 12, 5, 8, 7, 13
207	2A, 3A, 4D, 5D, 6G, 10HH	4, 6, 7, 6, 12, 6, 7, 7, 13
210	2A, 3A, 4D, 5D, 6G, 10AAA	6, 8, 7, 8, 10, 8, 7, 5, 13
256	2A, 3A, 4D, 5G, 6C, 10GGG	17, 13, 7, 4, 10, 15, 7, 2, 9
259	2A, 3A, 4D, 5G, 6G, 10Q	8, 4, 7, 3, 15, 10, 7, 3, 13
260*	2A, 3A, 4D, 5G, 6G, 10TT	7, 2, 8, 2, 16, 8, 8, 4, 13
291	2A, 3A, 4E, 5A, 6C, 10N	19, 16, 5, 2, 12, 1, 5, 6, 15
309	2A, 3A, 4E, 5B, 6C, 10R	18, 13, 0, 7, 7, 5, 0, 7, 15
310	2A, 3A, 4E, 5B, 6C, 10Z	15, 10, 0, 4, 10, 2, 0, 10, 15
311	2A, 3A, 4E, 5B, 6C, 10BB	14, 9, 0, 3, 11, 1, 0, 11, 15
312	2A, 3A, 4E, 5B, 6C, 10JJ	13, 8, 0, 2, 12, 0, 0, 12, 15
315	2A, 3A, 4E, 5B, 6G, 10R	14, 9, 0, 11, 7, 5, 0, 3, 19
316	2A, 3A, 4E, 5B, 6G, 10Z	11, 6, 0, 8, 10, 2, 0, 6, 19
317	2A, 3A, 4E, 5B, 6G, 10BB	10, 5, 0, 7, 11, 1, 0, 7, 19
318	2A, 3A, 4E, 5B, 6G, 10JJ	9, 4, 0, 6, 12, 0, 0, 8, 19
345	2A, 3A, 4E, 5D, 6C, 10HH	14, 10, 1, 2, 12, 0, 1, 11, 15
346	2A, 3A, 4E, 5D, 6C, 10YY	24, 20, 1, 12, 2, 10, 1, 1, 15
348	2A, 3A, 4E, 5D, 6C, 10AAA	16, 12, 1, 4, 10, 2, 1, 9, 15
351	2A, 3A, 4E, 5D, 6G, 10HH	10, 6, 1, 6, 12, 0, 1, 7, 19
354	2A, 3A, 4E, 5D, 6G, 10AAA	12, 8, 1, 8, 10, 2, 1, 5, 19
400	2A, 3A, 4E, 5G, 6C, 10GGG	23, 13, 1, 4, 10, 9, 1, 2, 15
403	2A, 3A, 4E, 5G, 6G, 10Q	14, 4, 1, 3, 15, 4, 1, 3, 19
404*	2A, 3A, 4E, 5G, 6G, 10TT	13, 2, 2, 2, 16, 2, 2, 4, 19
583	2A, 3B, 4A, 5A, 6O, 10E	15, 3, 27, 11, 7, 0, 27, 0, 8
584	2A, 3B, 4A, 5A, 6O, 10L	10, 8, 17, 6, 12, 5, 17, 5, 8
597	2A, 3B, 4A, 5B, 6F, 10R	7, 7, 8, 13, 7, 13, 8, 4, 10
598	2A, 3B, 4A, 5B, 6F, 10Z	4, 4, 8, 10, 10, 10, 8, 7, 10
599	2A, 3B, 4A, 5B, 6F, 10BB	3, 3, 8, 9, 11, 9, 8, 8, 10
600	2A, 3B, 4A, 5B, 6F, 10JJ	2, 2, 8, 8, 12, 8, 8, 9, 10
602	2A, 3B, 4A, 5B, 6O, 10P	10, 14, 4, 12, 6, 18, 4, 5, 8
603	2A, 3B, 4A, 5B, 6O, 10R	9, 9, 8, 11, 7, 13, 8, 6, 8

604	2A, 3B, 4A, 5B, 6O, 10Z	6, 6, 8, 8, 10, 10, 8, 9, 8
605	2A, 3B, 4A, 5B, 6O, 10BB	5, 5, 8, 7, 11, 9, 8, 10, 8
606	2A, 3B, 4A, 5B, 6O, 10JJ	4, 4, 8, 6, 12, 8, 8, 11, 8
613	2A, 3B, 4A, 5C, 6F, 10U	6, 10, 5, 14, 6, 15, 5, 5, 10
614	2A, 3B, 4A, 5C, 6F, 10KK	2, 4, 7, 10, 10, 9, 7, 9, 10
619	2A, 3B, 4A, 5C, 6O, 10U	8, 12, 5, 12, 6, 15, 5, 7, 8
620	2A, 3B, 4A, 5C, 6O, 10KK	4, 6, 7, 8, 10, 9, 7, 11, 8
631	2A, 3B, 4A, 5D, 6F, 10X	5, 1, 14, 10, 10, 5, 14, 6, 10
632	2A, 3B, 4A, 5D, 6F, 10DD	3, 3, 10, 8, 12, 7, 10, 8, 10
633	2A, 3B, 4A, 5D, 6F, 10HH	3, 4, 9, 8, 12, 8, 9, 8, 10
636	2A, 3B, 4A, 5D, 6F, 10AAA	5, 6, 9, 10, 10, 10, 9, 6, 10
637	2A, 3B, 4A, 5D, 6O, 10X	7, 3, 14, 8, 10, 5, 14, 8, 8
638	2A, 3B, 4A, 5D, 6O, 10DD	5, 5, 10, 6, 12, 7, 10, 10, 8
639	2A, 3B, 4A, 5D, 6O, 10HH	5, 6, 9, 6, 12, 8, 9, 10, 8
640	2A, 3B, 4A, 5D, 6O, 10YY	15, 16, 9, 16, 2, 18, 9, 0, 8
641	2A, 3B, 4A, 5D, 6O, 10ZZ	15, 0, 25, 16, 2, 2, 25, 0, 8
642	2A, 3B, 4A, 5D, 6O, 10AAA	7, 8, 9, 8, 10, 10, 9, 8, 8
656	2A, 3B, 4A, 5E, 6O, 10WW	15, 4, 24, 2, 16, 3, 24, 0, 8
686*	2A, 3B, 4A, 5G, 6F, 10TT	6, 0, 10, 4, 16, 10, 10, 5, 10
691	2A, 3B, 4A, 5G, 6O, 10Q	9, 4, 9, 3, 15, 12, 9, 6, 8
692*	2A, 3B, 4A, 5G, 6O, 10TT	8, 2, 10, 2, 16, 10, 10, 7, 6
694	2A, 3B, 4A, 5G, 6O, 10GGG	14, 9, 9, 8, 10, 17, 9, 1, 8
728	2A, 3B, 4D, 5A, 6O, 10L	12, 8, 15, 6, 12, 3, 15, 5, 10
741	2A, 3B, 4D, 5B, 6F, 10R	9, 7, 6, 13, 7, 11, 6, 4, 12
742	2A, 3B, 4D, 5B, 6F, 10Z	6, 4, 6, 10, 10, 8, 6, 7, 12
743	2A, 3B, 4D, 5B, 6F, 10BB	5, 3, 6, 9, 11, 7, 6, 8, 12
744	2A, 3B, 4D, 5B, 6F, 10JJ	4, 2, 6, 8, 12, 6, 6, 9, 12
746	2A, 3B, 4D, 5B, 6O, 10P	12, 14, 2, 12, 6, 16, 2, 5, 10
747	2A, 3B, 4D, 5B, 6O, 10R	11, 9, 6, 1, 1, 7, 11, 6, 6, 10
748	2A, 3B, 4D, 5B, 6O, 10Z	8, 6, 6, 8, 10, 8, 6, 9, 10
749	2A, 3B, 4D, 5B, 6O, 10BB	7, 5, 6, 7, 11, 7, 6, 10, 10
750	2A, 3B, 4D, 5B, 6O, 10JJ	6, 4, 6, 6, 12, 6, 6, 11, 10
757	2A, 3B, 4D, 5C, 6F, 10U	8, 10, 3, 14, 6, 13, 3, 5, 12
758	2A, 3B, 4D, 5C, 6F, 10KK	4, 4, 5, 10, 10, 7, 5, 9, 12
763	2A, 3B, 4D, 5C, 6O, 10U	10, 12, 3, 12, 6, 13, 3, 7, 10
764	2A, 3B, 4D, 5C, 6O, 10KK	6, 6, 5, 8, 10, 7, 5, 11, 10
775	2A, 3B, 4D, 5D, 6F, 10X	7, 1, 12, 10, 10, 3, 12, 6, 12
776	2A, 3B, 4D, 5D, 6F, 10DD	5, 3, 8, 8, 12, 5, 8, 8, 12
777	2A, 3B, 4D, 5D, 6F, 10HH	5, 4, 7, 8, 12, 6, 7, 8, 12
780	2A, 3B, 4D, 5D, 6F, 10AAA	7, 6, 7, 10, 10, 8, 7, 6, 12
781	2A, 3B, 4D, 5D, 6O, 10X	9, 3, 12, 8, 10, 3, 12, 8, 10
782	2A, 3B, 4D, 5D, 6O, 10DD	7, 5, 8, 6, 12, 5, 8, 10, 10
783	2A, 3B, 4D, 5D, 6O, 10HH	7, 6, 7, 6, 12, 6, 7, 10, 10
784	2A, 3B, 4D, 5D, 6O, 10YY	17, 16, 7, 16, 2, 16, 7, 0, 10
785	2A, 3B, 4D, 5D, 6O, 10ZZ	17, 0, 23, 16, 2, 0, 23, 0, 10

CONJUGACY OF Alt_5 AND $SL(2, 5)$ SUBGROUPS OF $E_8(\mathbb{C})$

786	2A, 3B, 4D, 5D, 6O, 10AAA	9, 8, 7, 8, 10, 8, 7, 8, 10
800	2A, 3B, 4D, 5E, 6O, 10WW	17, 4, 22, 2, 16, 1, 22, 0, 10
830*	2A, 3B, 4D, 5G, 6F, 10TT	8, 0, 8, 4, 16, 8, 8, 5, 12
835	2A, 3B, 4D, 5G, 6O, 10Q	11, 4, 7, 3, 15, 10, 7, 6, 10
836*	2A, 3B, 4D, 5G, 6O, 10TT	10, 2, 8, 2, 16, 8, 8, 7, 10
838	2A, 3B, 4D, 5G, 6O, 10GGG	16, 9, 7, 8, 10, 15, 7, 1, 10
885	2A, 3B, 4E, 5B, 6F, 10R	15, 7, 0, 13, 7, 5, 0, 4, 18
886	2A, 3B, 4E, 5B, 6F, 10Z	12, 4, 0, 1, 0, 10, 2, 0, 7, 18
887	2A, 3B, 4E, 5B, 6F, 10BB	11, 3, 0, 9, 11, 1, 0, 8, 18
888	2A, 3B, 4E, 5B, 6F, 10JJ	10, 2, 0, 8, 12, 0, 0, 9, 18
891	2A, 3B, 4E, 5B, 6O, 10R	17, 9, 0, 1, 1, 7, 5, 0, 6, 16
892	2A, 3B, 4E, 5B, 6O, 10Z	14, 6, 0, 8, 10, 2, 0, 9, 16
893	2A, 3B, 4E, 5B, 6O, 10BB	13, 5, 0, 7, 11, 1, 0, 10, 16
894	2A, 3B, 4E, 5B, 6O, 10JJ	12, 4, 0, 6, 12, 0, 0, 11, 16
921	2A, 3B, 4E, 5D, 6F, 10HH	11, 4, 1, 8, 12, 0, 1, 8, 18
924	2A, 3B, 4E, 5D, 6F, 10AAA	13, 6, 1, 1, 0, 10, 2, 1, 6, 18
927	2A, 3B, 4E, 5D, 6O, 10HH	13, 6, 1, 6, 12, 0, 1, 10, 16
928	2A, 3B, 4E, 5D, 6O, 10YY	23, 16, 1, 16, 2, 10, 1, 0, 16
930	2A, 3B, 4E, 5D, 6O, 10AAA	15, 8, 1, 8, 10, 2, 1, 8, 16
934	2A, 3B, 4E, 5D, 6P, 10YY	39, 32, 1, 0, 2, 10, 1, 16, 0
951	2A, 3B, 4E, 5E, 6P, 10BBB	55, 52, 0, 2, 0, 27, 0, 0, 0
953	2A, 3B, 4E, 5E, 6P, 10JJJ	54, 25, 26, 1, 1, 0, 26, 1, 0
974*	2A, 3B, 4E, 5G, 6F, 10TT	14, 0, 2, 4, 16, 2, 2, 5, 18
979	2A, 3B, 4E, 5G, 6O, 10Q	17, 4, 1, 3, 15, 4, 1, 6, 16
980*	2A, 3B, 4E, 5G, 6O, 10TT	16, 2, 2, 2, 16, 2, 2, 7, 16
982	2A, 3B, 4E, 5G, 6O, 10GGG	22, 9, 1, 8, 10, 9, 1, 1, 16
1153	2A, 3C, 4A, 5A, 6E, 10E	15, 0, 27, 14, 7, 0, 27, 0, 8
1159	2A, 3C, 4A, 5A, 6L, 10E	16, 1, 27, 13, 7, 0, 27, 1, 7
1173	2A, 3C, 4A, 5B, 6E, 10R	9, 6, 8, 14, 7, 13, 8, 6, 8
1174	2A, 3C, 4A, 5B, 6E, 10Z	6, 3, 8, 11, 10, 10, 8, 9, 8
1175	2A, 3C, 4A, 5B, 6E, 10BB	5, 2, 8, 10, 11, 9, 8, 10, 8
1176	2A, 3C, 4A, 5B, 6E, 10JJ	4, 1, 8, 9, 12, 8, 8, 11, 8
1179	2A, 3C, 4A, 5B, 6L, 10R	10, 7, 8, 13, 7, 13, 8, 7, 7
1180	2A, 3C, 4A, 5B, 6L, 10Z	7, 4, 8, 10, 10, 10, 8, 10, 7
1181	2A, 3C, 4A, 5B, 6L, 10BB	6, 3, 8, 9, 11, 9, 8, 11, 7
1182	2A, 3C, 4A, 5B, 6L, 10JJ	5, 2, 8, 8, 12, 8, 8, 12, 7
1190	2A, 3C, 4A, 5C, 6E, 10KK	4, 3, 7, 11, 10, 9, 7, 11, 8
1195	2A, 3C, 4A, 5C, 6L, 10U	9, 10, 5, 14, 6, 15, 5, 8, 7
1196	2A, 3C, 4A, 5C, 6L, 10KK	5, 4, 7, 10, 10, 9, 7, 12, 7
1207	2A, 3C, 4A, 5D, 6E, 10X	7, 0, 14, 11, 10, 5, 14, 8, 8
1208	2A, 3C, 4A, 5D, 6E, 10DD	5, 2, 10, 9, 12, 7, 10, 10, 8
1213	2A, 3C, 4A, 5D, 6L, 10X	8, 1, 14, 10, 10, 5, 14, 9, 7
1214	2A, 3C, 4A, 5D, 6L, 10DD	6, 3, 10, 8, 12, 7, 10, 11, 7
1215	2A, 3C, 4A, 5D, 6L, 10HH	6, 4, 9, 8, 12, 8, 9, 11, 7
1216	2A, 3C, 4A, 5D, 6L, 10YY	16, 14, 9, 18, 2, 18, 9, 1, 7

1218	2A, 3C, 4A, 5D, 6L, 10AAA	8, 6, 9, 10, 10, 10, 9, 9, 7
1268*	2A, 3C, 4A, 5G, 6L, 10TT	9, 0, 10, 4, 16, 10, 10, 8, 7
1310	2A, 3C, 4D, 5A, 6R, 10L	21, 14, 15, 0, 12, 3, 15, 14, 1
1311	2A, 3C, 4D, 5A, 6R, 10N	21, 18, 11, 0, 12, 7, 11, 14, 1
1317	2A, 3C, 4D, 5B, 6E, 10R	11, 6, 6, 14, 7, 11, 6, 6, 10
1318	2A, 3C, 4D, 5B, 6E, 10Z	8, 3, 6, 11, 10, 8, 6, 9, 10
1319	2A, 3C, 4D, 5B, 6E, 10BB	7, 2, 6, 10, 11, 7, 6, 10, 10
1320	2A, 3C, 4D, 5B, 6E, 10JJ	6, 1, 6, 9, 12, 6, 6, 11, 10
1323	2A, 3C, 4D, 5B, 6L, 10R	12, 7, 6, 13, 7, 11, 6, 7, 9
1324	2A, 3C, 4D, 5B, 6L, 10Z	9, 4, 6, 10, 10, 8, 6, 10, 9
1325	2A, 3C, 4D, 5B, 6L, 10BB	8, 3, 6, 9, 11, 7, 6, 11, 9
1326	2A, 3C, 4D, 5B, 6L, 10JJ	7, 2, 6, 8, 12, 6, 6, 12, 9
1328	2A, 3C, 4D, 5B, 6R, 10P	21, 20, 2, 6, 6, 16, 2, 14, 1
1329	2A, 3C, 4D, 5B, 6R, 10R	20, 15, 6, 5, 7, 11, 6, 15, 1
1330	2A, 3C, 4D, 5B, 6R, 10Z	17, 12, 6, 2, 10, 8, 6, 18, 1
1331	2A, 3C, 4D, 5B, 6R, 10BB	16, 11, 6, 1, 11, 7, 6, 19, 1
1332	2A, 3C, 4D, 5B, 6R, 10JJ	15, 10, 6, 0, 12, 6, 6, 20, 1
1334	2A, 3C, 4D, 5C, 6E, 10KK	6, 3, 5, 11, 10, 7, 5, 11, 10
1339	2A, 3C, 4D, 5C, 6L, 10U	11, 10, 3, 14, 6, 13, 3, 8, 9
1340	2A, 3C, 4D, 5C, 6L, 10KK	7, 4, 5, 10, 10, 7, 5, 12, 9
1345	2A, 3C, 4D, 5C, 6R, 10U	19, 18, 3, 6, 6, 13, 3, 16, 1
1346	2A, 3C, 4D, 5C, 6R, 10KK	15, 12, 5, 2, 10, 7, 5, 20, 1
1351	2A, 3C, 4D, 5D, 6E, 10X	9, 0, 12, 11, 10, 3, 12, 8, 10
1352	2A, 3C, 4D, 5D, 6E, 10DD	7, 2, 8, 9, 12, 5, 8, 10, 10
1357	2A, 3C, 4D, 5D, 6L, 10X	10, 1, 12, 10, 10, 3, 12, 9, 9
1358	2A, 3C, 4D, 5D, 6L, 10DD	8, 3, 8, 8, 12, 5, 8, 11, 9
1359	2A, 3C, 4D, 5D, 6L, 10HH	8, 4, 7, 8, 12, 6, 7, 11, 9
1360	2A, 3C, 4D, 5D, 6L, 10YY	18, 14, 7, 18, 2, 16, 7, 1, 9
1362	2A, 3C, 4D, 5D, 6L, 10AAA	10, 6, 7, 1, 0, 10, 8, 7, 9, 9
1363	2A, 3C, 4D, 5D, 6R, 10X	18, 9, 12, 2, 10, 3, 12, 17, 1
1364	2A, 3C, 4D, 5D, 6R, 10DD	16, 11, 8, 0, 12, 5, 8, 19, 1
1365	2A, 3C, 4D, 5D, 6R, 10HH	16, 12, 7, 0, 12, 6, 7, 19, 1
1366	2A, 3C, 4D, 5D, 6R, 10YY	26, 22, 7, 10, 2, 16, 7, 9, 1
1367	2A, 3C, 4D, 5D, 6R, 10ZZ	26, 6, 23, 10, 2, 0, 23, 9, 1
1368	2A, 3C, 4D, 5D, 6R, 10AAA	18, 14, 7, 2, 10, 8, 7, 17, 1
1401	2A, 3C, 4D, 5F, 6R, 10EEE	35, 20, 31, 0, 12, 1, 31, 0, 1
1412*	2A, 3C, 4D, 5G, 6L, 10TT	11, 0, 8, 4, 16, 8, 8, 8, 9
1419	2A, 3C, 4D, 5G, 6R, 10DDD	35, 32, 0, 12, 0, 32, 0, 0, 1
1420	2A, 3C, 4D, 5G, 6R, 10GGG	25, 15, 7, 2, 10, 15, 7, 10, 1
1455	2A, 3C, 4E, 5A, 6R, 10N	27, 18, 5, 0, 12, 1, 5, 14, 7
1461	2A, 3C, 4E, 5B, 6E, 10R	17, 6, 0, 14, 7, 5, 0, 6, 16
1462	2A, 3C, 4E, 5B, 6E, 10Z	14, 3, 0, 11, 10, 2, 0, 9, 16
1463	2A, 3C, 4E, 5B, 6E, 10BB	13, 2, 0, 10, 11, 1, 0, 10, 16
1464	2A, 3C, 4E, 5B, 6E, 10JJ	12, 1, 0, 9, 12, 0, 0, 11, 16
1467	2A, 3C, 4E, 5B, 6L, 10R	18, 7, 0, 13, 7, 5, 0, 7, 15

1468	2A, 3C, 4E, 5B, 6L, 10Z	15, 4, 0, 10, 10, 2, 0, 10, 15
1469	2A, 3C, 4E, 5B, 6L, 10BB	14, 3, 0, 9, 11, 1, 0, 11, 15
1470	2A, 3C, 4E, 5B, 6L, 10JJ	13, 2, 0, 8, 12, 0, 0, 12, 15
1473	2A, 3C, 4E, 5B, 6R, 10R	26, 15, 0, 5, 7, 5, 0, 15, 7
1474	2A, 3C, 4E, 5B, 6R, 10Z	23, 12, 0, 2, 10, 2, 0, 18, 7
1475	2A, 3C, 4E, 5B, 6R, 10BB	22, 11, 0, 1, 11, 1, 0, 19, 7
1476	2A, 3C, 4E, 5B, 6R, 10JJ	21, 10, 0, 0, 12, 0, 0, 20, 7
1503	2A, 3C, 4E, 5D, 6L, 10HH	14, 4, 1, 8, 12, 0, 1, 11, 15
1504	2A, 3C, 4E, 5D, 6L, 10YY	24, 14, 1, 18, 2, 10, 1, 1, 15
1506	2A, 3C, 4E, 5D, 6L, 10AAA	16, 6, 1, 10, 10, 2, 1, 9, 15
1509	2A, 3C, 4E, 5D, 6R, 10HH	22, 12, 1, 0, 12, 0, 1, 19, 7
1510	2A, 3C, 4E, 5D, 6R, 10YY	32, 22, 1, 10, 2, 10, 1, 9, 7
1512	2A, 3C, 4E, 5D, 6R, 10AAA	24, 14, 1, 2, 10, 2, 1, 17, 7
1556*	2A, 3C, 4E, 5G, 6L, 10TT	17, 0, 2, 4, 16, 2, 2, 8, 15
1564	2A, 3C, 4E, 5G, 6R, 10GGG	31, 15, 1, 2, 10, 9, 1, 10, 7
2105	2A, 3D, 4E, 5E, 6T, 10JJJ	53, 0, 26, 26, 1, 0, 26, 0, 1
2294	2A, 3D, 4G, 5H, 6S, 10CCC	133, 56, 0, 0, 0, 1, 0, 0, 0
2305	2B, 3A, 4B, 5A, 6A, 10A	36, 48, 0, 8, 0, 28, 0, 0, 0
2306	2B, 3A, 4B, 5A, 6A, 10F	29, 27, 14, 1, 7, 7, 14, 7, 0
2314	2B, 3A, 4B, 5A, 6I, 10EE	8, 8, 12, 4, 16, 0, 12, 4, 12
2323	2B, 3A, 4B, 5B, 6A, 10I	29, 34, 0, 7, 1, 21, 0, 7, 0
2324	2B, 3A, 4B, 5B, 6A, 10O	24, 28, 1, 2, 6, 15, 1, 12, 0
2330	2B, 3A, 4B, 5B, 6I, 10O	12, 16, 1, 14, 6, 15, 1, 0, 12
2331	2B, 3A, 4B, 5B, 6I, 10AA	8, 10, 3, 10, 10, 9, 3, 4, 12
2332	2B, 3A, 4B, 5B, 6I, 10II	6, 8, 3, 8, 12, 7, 3, 6, 12
2333	2B, 3A, 4B, 5B, 6I, 10LL	5, 6, 4, 7, 13, 5, 4, 7, 12
2342*	2B, 3A, 4B, 5C, 6A, 10FF	20, 20, 6, 0, 8, 6, 6, 16, 0
2348*	2B, 3A, 4B, 5C, 6I, 10FF	8, 8, 6, 12, 8, 6, 6, 4, 12
2349*	2B, 3A, 4B, 5C, 6I, 10OO	4, 6, 4, 8, 12, 4, 4, 8, 12
2363	2B, 3A, 4B, 5D, 6A, 10QQ	23, 16, 14, 0, 8, 1, 14, 13, 0
2365	2B, 3A, 4B, 5D, 6I, 10V	9, 13, 3, 10, 10, 10, 3, 3, 12
2366	2B, 3A, 4B, 5D, 6I, 10CC	7, 7, 7, 8, 12, 4, 7, 5, 12
2367	2B, 3A, 4B, 5D, 6I, 10MM	5, 6, 6, 6, 14, 3, 6, 7, 12
2369	2B, 3A, 4B, 5D, 6I, 10QQ	11, 4, 14, 12, 8, 1, 14, 1, 12
2419	2B, 3A, 4B, 5G, 6I, 10S	11, 7, 5, 5, 15, 10, 5, 1, 12
2420*	2B, 3A, 4B, 5G, 6I, 10GG	8, 3, 6, 2, 18, 6, 6, 4, 12
2421*	2B, 3A, 4B, 5G, 6I, 10SS	10, 4, 7, 4, 16, 7, 7, 2, 12
2458	2B, 3A, 4C, 5A, 6I, 10EE	6, 8, 14, 4, 16, 2, 14, 4, 10
2474	2B, 3A, 4C, 5B, 6I, 10O	10, 16, 3, 14, 6, 17, 3, 0, 10
2475	2B, 3A, 4C, 5B, 6I, 10AA	6, 10, 5, 10, 10, 11, 5, 4, 10
2476	2B, 3A, 4C, 5B, 6I, 10II	4, 8, 5, 8, 12, 9, 5, 6, 10
2477	2B, 3A, 4C, 5B, 6I, 10LL	3, 6, 6, 7, 13, 7, 6, 7, 10
2491	2B, 3A, 4C, 5C, 6I, 10T	10, 20, 0, 16, 4, 20, 0, 0, 10
2492*	2B, 3A, 4C, 5C, 6I, 10FF	6, 8, 8, 12, 8, 8, 8, 4, 10

2493*	2B, 3A, 4C, 5C, 6I, 10OO	2, 6, 6, 8, 12, 6, 6, 8, 10
2509	2B, 3A, 4C, 5D, 6I, 10V	7, 13, 5, 10, 10, 12, 5, 3, 10
2510	2B, 3A, 4C, 5D, 6I, 10CC	5, 7, 9, 8, 12, 6, 9, 5, 10
2511	2B, 3A, 4C, 5D, 6I, 10MM	3, 6, 8, 6, 14, 5, 8, 7, 10
2513	2B, 3A, 4C, 5D, 6I, 10QQ	9, 4, 16, 12, 8, 3, 16, 1, 10
2563	2B, 3A, 4C, 5G, 6I, 10S	9, 7, 7, 5, 15, 12, 7, 1, 10
2564*	2B, 3A, 4C, 5G, 6I, 10GG	6, 3, 8, 2, 18, 8, 8, 4, 10
2565*	2B, 3A, 4C, 5G, 6I, 10SS	8, 4, 9, 4, 16, 9, 9, 2, 10
2888	2B, 3B, 4B, 5A, 6K, 10F	24, 19, 14, 9, 7, 7, 14, 2, 5
2889	2B, 3B, 4B, 5A, 6K, 10Y	16, 15, 10, 1, 15, 3, 10, 10, 5
2890	2B, 3B, 4B, 5A, 6K, 10EE	15, 12, 12, 0, 16, 0, 12, 11, 5
2900	2B, 3B, 4B, 5B, 6H, 10O	13, 14, 1, 16, 6, 15, 1, 1, 11
2901	2B, 3B, 4B, 5B, 6H, 10AA	9, 8, 3, 12, 10, 9, 3, 5, 11
2902	2B, 3B, 4B, 5B, 6H, 10II	7, 6, 3, 10, 12, 7, 3, 7, 11
2903	2B, 3B, 4B, 5B, 6H, 10LL	6, 4, 4, 9, 13, 5, 4, 8, 11
2905	2B, 3B, 4B, 5B, 6K, 10I	24, 26, 0, 15, 1, 21, 0, 2, 5
2906	2B, 3B, 4B, 5B, 6K, 10O	19, 20, 1, 10, 6, 15, 1, 7, 5
2907	2B, 3B, 4B, 5B, 6K, 10AA	15, 14, 3, 6, 10, 9, 3, 11, 5
2908	2B, 3B, 4B, 5B, 6K, 10II	13, 12, 3, 4, 12, 7, 3, 13, 5
2909	2B, 3B, 4B, 5B, 6K, 10LL	12, 10, 4, 3, 13, 5, 4, 14, 5
2918*	2B, 3B, 4B, 5C, 6H, 10FF	9, 6, 6, 14, 8, 6, 6, 5, 11
2919*	2B, 3B, 4B, 5C, 6H, 10OO	5, 4, 4, 10, 12, 4, 4, 9, 11
2924*	2B, 3B, 4B, 5C, 6K, 10FF	15, 12, 6, 8, 8, 6, 6, 11, 5
2925*	2B, 3B, 4B, 5C, 6K, 10OO	11, 10, 4, 4, 12, 4, 4, 15, 5
2936	2B, 3B, 4B, 5D, 6H, 10CC	8, 5, 7, 10, 12, 4, 7, 6, 11
2937	2B, 3B, 4B, 5D, 6H, 10MM	6, 4, 6, 8, 14, 3, 6, 8, 11
2939	2B, 3B, 4B, 5D, 6H, 10QQ	12, 2, 14, 14, 8, 1, 14, 2, 11
2941	2B, 3B, 4B, 5D, 6K, 10V	16, 17, 3, 6, 10, 10, 3, 10, 5
2942	2B, 3B, 4B, 5D, 6K, 10CC	14, 11, 7, 4, 12, 4, 7, 12, 5
2943	2B, 3B, 4B, 5D, 6K, 10MM	12, 10, 6, 2, 14, 3, 6, 14, 5
2945	2B, 3B, 4B, 5D, 6K, 10QQ	18, 8, 14, 8, 8, 1, 14, 8, 5
2961	2B, 3B, 4B, 5E, 6K, 10XX	24, 16, 15, 0, 16, 6, 15, 2, 5
2989	2B, 3B, 4B, 5G, 6H, 10S	12, 5, 5, 7, 15, 10, 5, 2, 11
2990*	2B, 3B, 4B, 5G, 6H, 10GG	9, 1, 6, 4, 18, 6, 6, 5, 11
2991*	2B, 3B, 4B, 5G, 6H, 10SS	11, 2, 7, 6, 16, 7, 7, 3, 11
2995	2B, 3B, 4B, 5G, 6K, 10S	18, 11, 5, 1, 15, 10, 5, 8, 5
2997*	2B, 3B, 4B, 5G, 6K, 10SS	17, 8, 7, 0, 16, 7, 7, 9, 5
3032	2B, 3B, 4C, 5A, 6K, 10F	22, 19, 16, 9, 7, 9, 16, 2, 3
3033	2B, 3B, 4C, 5A, 6K, 10Y	14, 15, 12, 1, 15, 5, 12, 10, 3
3034	2B, 3B, 4C, 5A, 6K, 10EE	13, 12, 14, 0, 16, 2, 14, 11, 3
3044	2B, 3B, 4C, 5B, 6H, 10O	11, 14, 3, 16, 6, 17, 3, 1, 9
3045	2B, 3B, 4C, 5B, 6H, 10AA	7, 8, 5, 12, 10, 11, 5, 5, 9
3046	2B, 3B, 4C, 5B, 6H, 10II	5, 6, 5, 10, 12, 9, 5, 7, 9
3047	2B, 3B, 4C, 5B, 6H, 10LL	4, 4, 6, 9, 13, 7, 6, 8, 9

3049	2B, 3B, 4C, 5B, 6K, 10I	22, 26, 2, 15, 1, 23, 2, 2, 3
3050	2B, 3B, 4C, 5B, 6K, 10O	17, 20, 3, 10, 6, 17, 3, 7, 3
3051	2B, 3B, 4C, 5B, 6K, 10AA	13, 14, 5, 6, 10, 11, 5, 11, 3
3052	2B, 3B, 4C, 5B, 6K, 10II	11, 12, 5, 4, 12, 9, 5, 13, 3
3053	2B, 3B, 4C, 5B, 6K, 10LL	10, 10, 6, 3, 13, 7, 6, 14, 3
3062*	2B, 3B, 4C, 5C, 6H, 10FF	7, 6, 8, 14, 8, 8, 8, 5, 9
3063*	2B, 3B, 4C, 5C, 6H, 10OO	3, 4, 6, 10, 12, 6, 6, 9, 9
3067	2B, 3B, 4C, 5C, 6K, 10T	17, 24, 0, 12, 4, 20, 0, 7, 3
3068*	2B, 3B, 4C, 5C, 6K, 10FF	13, 12, 8, 8, 8, 8, 8, 11, 3
3069*	2B, 3B, 4C, 5C, 6K, 10OO	9, 10, 6, 4, 12, 6, 6, 15, 3
3080	2B, 3B, 4C, 5D, 6H, 10CC	6, 5, 9, 10, 12, 6, 9, 6, 9
3081	2B, 3B, 4C, 5D, 6H, 10MM	4, 4, 8, 8, 14, 5, 8, 8, 9
3083	2B, 3B, 4C, 5D, 6H, 10QQ	10, 2, 16, 14, 8, 3, 16, 2, 9
3085	2B, 3B, 4C, 5D, 6K, 10V	14, 17, 5, 6, 10, 12, 5, 10, 3
3086	2B, 3B, 4C, 5D, 6K, 10CC	12, 11, 9, 4, 12, 6, 9, 12, 3
3087	2B, 3B, 4C, 5D, 6K, 10MM	10, 10, 8, 2, 14, 5, 8, 14, 3
3088	2B, 3B, 4C, 5D, 6K, 10PP	24, 32, 0, 16, 0, 27, 0, 0, 3
3089	2B, 3B, 4C, 5D, 6K, 10QQ	16, 8, 16, 8, 8, 3, 16, 8, 3
3105	2B, 3B, 4C, 5E, 6K, 10XX	22, 16, 17, 0, 16, 8, 17, 2, 3
3133	2B, 3B, 4C, 5G, 6H, 10S	10, 5, 7, 7, 15, 12, 7, 2, 9
3134*	2B, 3B, 4C, 5G, 6H, 10GG	7, 1, 8, 4, 18, 8, 8, 5, 9
3135*	2B, 3B, 4C, 5G, 6H, 10SS	9, 2, 9, 6, 16, 9, 9, 3, 9
3139	2B, 3B, 4C, 5G, 6K, 10S	16, 11, 7, 1, 15, 12, 7, 8, 3
3141*	2B, 3B, 4C, 5G, 6K, 10SS	15, 8, 9, 0, 16, 9, 9, 9, 3
3458	2B, 3C, 4B, 5A, 6D, 10F	22, 14, 14, 14, 7, 7, 14, 0, 7
3475	2B, 3C, 4B, 5B, 6D, 10I	22, 21, 0, 20, 1, 21, 0, 0, 7
3476	2B, 3C, 4B, 5B, 6D, 10O	17, 15, 1, 15, 6, 15, 1, 5, 7
3477	2B, 3C, 4B, 5B, 6D, 10AA	13, 9, 3, 11, 10, 9, 3, 9, 7
3478	2B, 3C, 4B, 5B, 6D, 10II	11, 7, 3, 9, 12, 7, 3, 11, 7
3479	2B, 3C, 4B, 5B, 6D, 10LL	10, 5, 4, 8, 13, 5, 4, 12, 7
3483	2B, 3C, 4B, 5B, 6J, 10AA	10, 6, 3, 14, 10, 9, 3, 6, 10
3484	2B, 3C, 4B, 5B, 6J, 10II	8, 4, 3, 12, 12, 7, 3, 8, 10
3485	2B, 3C, 4B, 5B, 6J, 10LL	7, 2, 4, 11, 13, 5, 4, 9, 10
3494*	2B, 3C, 4B, 5C, 6D, 10FF	13, 7, 6, 13, 8, 6, 6, 9, 7
3495*	2B, 3C, 4B, 5C, 6D, 10OO	9, 5, 4, 9, 12, 4, 4, 13, 7
3500*	2B, 3C, 4B, 5C, 6J, 10FF	10, 4, 6, 16, 8, 6, 6, 6, 10
3501*	2B, 3C, 4B, 5C, 6J, 10OO	6, 2, 4, 12, 12, 4, 4, 10, 10
3511	2B, 3C, 4B, 5D, 6D, 10V	14, 12, 3, 11, 10, 10, 3, 8, 7
3512	2B, 3C, 4B, 5D, 6D, 10CC	12, 6, 7, 9, 12, 4, 7, 10, 7
3513	2B, 3C, 4B, 5D, 6D, 10MM	10, 5, 6, 7, 14, 3, 6, 12, 7
3515	2B, 3C, 4B, 5D, 6D, 10QQ	16, 3, 14, 13, 8, 1, 14, 6, 7
3518	2B, 3C, 4B, 5D, 6J, 10CC	9, 3, 7, 12, 12, 4, 7, 7, 10
3519	2B, 3C, 4B, 5D, 6J, 10MM	7, 2, 6, 10, 14, 3, 6, 9, 10

3521	2B, 3C, 4B, 5D, 6J, 10QQ	13, 0, 14, 16, 8, 1, 14, 3, 10
3565	2B, 3C, 4B, 5G, 6D, 10S	16, 6, 5, 6, 15, 10, 5, 6, 7
3566*	2B, 3C, 4B, 5G, 6D, 10GG	13, 2, 6, 3, 18, 6, 6, 9, 7
3567*	2B, 3C, 4B, 5G, 6D, 10SS	15, 3, 7, 5, 16, 7, 7, 7, 7
3573*	2B, 3C, 4B, 5G, 6J, 10SS	12, 0, 7, 8, 16, 7, 7, 4, 10
3602	2B, 3C, 4C, 5A, 6D, 10F	20, 14, 16, 14, 7, 9, 16, 0, 5
3619	2B, 3C, 4C, 5B, 6D, 10I	20, 21, 2, 20, 1, 23, 2, 0, 5
3620	2B, 3C, 4C, 5B, 6D, 10O	15, 15, 3, 15, 6, 17, 3, 5, 5
3621	2B, 3C, 4C, 5B, 6D, 10AA	11, 9, 5, 11, 10, 11, 5, 9, 5
3622	2B, 3C, 4C, 5B, 6D, 10II	9, 7, 5, 9, 12, 9, 5, 11, 5
3623	2B, 3C, 4C, 5B, 6D, 10LL	8, 5, 6, 8, 13, 7, 6, 12, 5
3627	2B, 3C, 4C, 5B, 6J, 10AA	8, 6, 5, 14, 10, 11, 5, 6, 8
3628	2B, 3C, 4C, 5B, 6J, 10II	6, 4, 5, 12, 12, 9, 5, 8, 8
3629	2B, 3C, 4C, 5B, 6J, 10LL	5, 2, 6, 11, 13, 7, 6, 9, 8
3638*	2B, 3C, 4C, 5C, 6D, 10FF	11, 7, 8, 13, 8, 8, 8, 9, 5
3639*	2B, 3C, 4C, 5C, 6D, 10OO	7, 5, 6, 9, 12, 6, 6, 13, 5
3644*	2B, 3C, 4C, 5C, 6J, 10FF	8, 4, 8, 16, 8, 8, 8, 6, 8
3645*	2B, 3C, 4C, 5C, 6J, 10OO	4, 2, 6, 12, 12, 6, 6, 10, 8
3655	2B, 3C, 4C, 5D, 6D, 10V	12, 12, 5, 11, 10, 12, 5, 8, 5
3656	2B, 3C, 4C, 5D, 6D, 10CC	10, 6, 9, 9, 12, 6, 9, 10, 5
3657	2B, 3C, 4C, 5D, 6D, 10MM	8, 5, 8, 7, 14, 5, 8, 12, 5
3659	2B, 3C, 4C, 5D, 6D, 10QQ	14, 3, 16, 13, 8, 3, 16, 6, 5
3662	2B, 3C, 4C, 5D, 6J, 10CC	7, 3, 9, 12, 12, 6, 9, 7, 8
3663	2B, 3C, 4C, 5D, 6J, 10MM	5, 2, 8, 10, 14, 5, 8, 9, 8
3665	2B, 3C, 4C, 5D, 6J, 10QQ	11, 0, 16, 16, 8, 3, 16, 3, 8
3709	2B, 3C, 4C, 5G, 6D, 10S	14, 6, 7, 6, 15, 12, 7, 6, 5
3710*	2B, 3C, 4C, 5G, 6D, 10GG	11, 2, 8, 3, 18, 8, 8, 9, 5
3711*	2B, 3C, 4C, 5G, 6D, 10SS	13, 3, 9, 5, 16, 9, 9, 7, 5
3717*	2B, 3C, 4C, 5G, 6J, 10SS	10, 0, 9, 8, 16, 9, 9, 4, 8
3847	2B, 3C, 4F, 5F, 6Q, 10B	78, 64, 0, 0, 0, 14, 0, 0, 0
3868*	2B, 3C, 4F, 5G, 6Q, 10FFF	66, 32, 1, 0, 0, 1, 1, 12, 0
4438*	2B, 3D, 4F, 5G, 6M, 10FFF	55, 0, 1, 32, 0, 1, 1, 1, 11

Notation 4.10 For ease of writing, we do not use the subscripts in the Atlas notation from here on. To distinguish the two characters of degree three (resp. two) we use 3 (resp. 2) to denote the character 3_a (resp. 2_a) and 3' (resp. 2') to denote the character 3_b (resp. 2_b). To distinguish between the faithful and nonfaithful characters of SL(2, 5) of degree four, we use 4_f to denote the

faithful character and 4 to denote the nonfaithful character (which is also a character of Alt$_5$).

Remark 4.11 In what follows, we will be making lists of embeddings of Alt$_5$ and SL(2, 5)-subgroups of various subgroups of G. Define ψ : {n-dimensional characters of SL(2, 5)} $\to Z^9$ be the map defined by $\psi(\eta) = ((\eta, 1), (\eta, 3), (\eta, 3'), (\eta, 4), (\eta, 5), (\eta, 2), (\eta, 2'), (\eta, 4_f), (\eta, 6))$. Note that ψ is injective so ψ^{-1} is well-defined. Let $\mu : Z^9 \to Z^9$ be defined by $\mu(a, b, c, d, e, f, g, h, i) = (a, c, b, d, e, g, f, h, i)$. Then $\upsilon := \psi^{-1}(\mu(\psi(\eta)))$ is the character of SL(2, 5) in which the multiplicities of 3 and 3' for η and the multiplicities of 2 and 2' for η are interchanged. If y is an element of order 10 in M $\cong$ SL(2, 5), then $\eta(y) = \upsilon(y^3)$. If y is an element of order 2, 3, 4 or 6 in M then $\eta(y) = \upsilon(y)$. Hence, the affording representations for η and υ are two nonconjugate embeddings of M into GL(n, $\mathbb{C}$). We call μ an *outer twist* and say υ is obtained from η by an outer twist. At times we will abuse notation and refer to an outer twist of an equivalence class of embeddings of Alt$_5$ or SL(2, 5) into GL(n, $\mathbb{C}$). By this we simply mean that the classes of embeddings affording η and υ are interchanged.

Lemma 4.12 Suppose L_1 and L_2 are two Alt$_5$ (resp. SL(2, 5)) subgroups of an odd quotient Q of $S := \prod_{i=1}^{n} SL(2j+1, \mathbb{C})$ where j is some fixed integer . Let $\alpha : S \to Q$ be the natural projection. Then $\alpha^{-1}(L_1)$ and $\alpha^{-1}(L_2)$ each contain a unique Alt$_5$ (resp. SL(2, 5)) subgroup. Moreover, if A_i is the unique Alt$_5$ (resp. SL(2, 5) subgroup of $\alpha^{-1}(L_i)$ for i = 1, 2, then A_1 is conjugate in S to A_2 if and only if L_1 is conjugate in Q to L_2.

Proof. Since L_1 and L_2 are perfect, they have unique covering groups by [Hup '67](V.23.6). By [Hup '67](V.25.7), the Schur multiplier of Alt_5 is Z_2. Hence, the extension kernel of any odd cyclic extension E of L_1 or L_2 intersects E' trivially, and therefore by Lemma 1.38, since $ker\alpha$ has odd order, $\alpha^{-1}(L_i) \cong ker\alpha \times L_i$. Since Alt_5 (resp. SL(2, 5)) is perfect, any Alt_5 (resp. SL(2, 5)) subgroup of $\alpha^{-1}(L_i)$ is contained in $\alpha^{-1}(L_i)'$. But $\alpha^{-1}(L_i)' \cong Alt_5$ (resp. SL(2, 5)) so is the unique Alt_5 (resp. SL(2, 5)) subgroup of $\alpha^{-1}(L_i)$, and so $A_i := \alpha^{-1}(L_i)'$.

Now suppose L_1 is conjugate in Q to L_2, say $(L_1)^g = L_2$. Let $x \in \alpha^{-1}(g)$. Then $(A_1)^x \cong Alt_5$ (resp. SL(2, 5)) and is a subgroup of $\alpha^{-1}(L_2)$. Hence $(A_1)^x = A_2$. The other direction of the if and only if statement is trivial. ∎

Lemma 4.13 Two subgroups A_1, A_2 of S (where S is as in Lemma 4.12) are conjugate in S if and only if they are conjugate in $P := \prod_{i=1}^{n} GL(2j+1, \mathbb{C})$.

Proof. Suppose $(A_1)^g = A_2$ where $g \in P$. Say $g = g_1 ... g_n$, where g_i is an element of the ith factor of P. Let $a = diag\left(^{(2j+1)}\sqrt{1/det(g_1)}...^{(2j+1)}\sqrt{1/det(g_1)}\right) \times ... \times diag\left(^{(2j+1)}\sqrt{1/det(g_n)}...^{(2j+1)}\sqrt{1/det(g_n)}\right) \in P$. (Here we mean the real $(2j+1)^{th}$ root of $1/det(g_i)$ which is unique for each i since $2j+1$ is odd). Then $det(a) = 1/det(g)$, so $ga \in S$. Since $a \in Z(P)$, $(A_1)^{ga} = A_2$. The converse is trivial. ∎

Remark 4.14 Using the preceding Lemmas with $n = 1$ and $j = 4$, we can find the number of $\mathcal{A}$-classes of Alt_5 (resp. SL(2, 5)) subgroups of $\mathcal{A}$ by looking at 9-dimensional faithful representations of Alt_5 (resp. SL(2, 5)). Since Alt_5 (resp. SL(2, 5)) is perfect, every Alt_5 (resp. SL(2, 5)) subgroup of $GL(9, \mathbb{C})$ is contained in $SL(9, \mathbb{C})$. Using representation theory, we know that two Alt_5 (resp. SL(2, 5)) subgroups of $GL(9, \mathbb{C})$ (and therefore of $SL(9, \mathbb{C})$) are conjugate in $GL(9, \mathbb{C})$ if and only if the 9-dimensional modules they act on are isomorphic. By Lemma 4.13, they are conjugate in GL(9, $\mathbb{C}$) if and only if they are conjugate in SL(9, $\mathbb{C}$). Hence, by Lemma 4.12, we can determine whether two Alt_5 (resp. SL(2, 5)) subgroups of $\mathcal{A}$ are conjugate in $\mathcal{A}$.

Notation 4.15 In what follows, when we use the term equivalence class of embeddings (into a subgroup of GL(n, $\mathbb{C}$)), we mean a class of embeddings which are equivalent as representations in GL(n, $\mathbb{C}$) where n is the degree of the embedding.

Table 4.16 Conjugacy classes of Alt_5 subgroups of $\mathcal{A}$.

The following is a list of all 9-dimensional faithful characters of Alt_5 modulo an outer twist (see Remark 4.11) and therefore is a list of all SL(9, $\mathbb{C}$)-classes of Alt_5 subgroups of SL(9, $\mathbb{C}$). (An outer twist switches the multiplicities of the constituents 3 and 3' which affords a fusion pattern which is in the same equivalence class as the original fusion pattern in the sense of Remark 4.8). Each class of subgroups corresponds (by α) to a unique $\mathcal{A}$-class of Alt_5 subgroups of $\mathcal{A}$ whose fusion pattern in G is calculated using the formula $\chi|_\mathcal{A} = \text{adjoint}(sl_9(V)) + \Lambda^3(V) + \Lambda^3(V^*)$ (see [FuHa '91] p. 361), where V is a natural nine-dimensional module for $\mathcal{A}$, and is given in the second column.

Representation	Corresponding fusion pattern	
$3 + 1^6$	769	2A, 3D, 5H
$4 + 1^5$	753	2A, 3D, 5G
$5 + 1^4$	557	2A, 3C, 5G
$3^2 + 1^3$	1312	2B, 3C, 5F
$3+3'+1^3$	1341	2B, 3C, 5G
$4 + 3 + 1^2$	1252	2B, 3C, 5D
$4^2 + 1$	1236	2B, 3C, 5C
$5 + 3 + 1$	1056	2B, 3B, 5D

3^3	302	2A, 3B, 5E
$3^2 + 3'$	272	2A, 3B, 5D
$5 + 4$	1040	2B, 3B, 5C

Remark 4.17 Since Table 4.16 shows that every representation affords a different fusion pattern, we see that any two Alt_5 subgroups of $\mathcal{A}$ which are not conjugate in $\mathcal{A}$ are not conjugate in G. Hence Table 4.16 may be regarded as a list of the G-conjugacy classes of Alt_5 subgroups of $\mathcal{A}$.

Table 4.18 Conjugacy classes of SL(2, 5) subgroups of $\mathcal{A}$.

The following is a list of all 9-dimensional faithful characters of SL(2, 5) modulo an outer twist (see Remark 4.11) and is therefore a list of all SL(9, $\mathbb{C}$)-classes of SL(2, 5) subgroups of SL(9, $\mathbb{C}$) (an outer twist switches the multiplicities of the constituents 3 and 3' and 2 and 2' which affords a fusion pattern which is in the same equivalence class as the original fusion pattern in the sense of Remark 4.8). Each class of subgroups corresponds (by α) to a unique $\mathcal{A}$-class of SL(2, 5) subgroups of $\mathcal{A}$ whose fusion pattern in G is calculated using the formula $\chi|_\mathcal{A} = \text{adjoint}(sl_9(V)) + \Lambda^3(V) + \Lambda^3(V^*)$ (see [FuHa '91] p. 361), where V is a natural nine-dimensional module for $\mathcal{A}$, and is given in the second column.

Representation	Corresponding fusion pattern	
$2 + 1^7$	2294	2A, 4G, 3D, 5H, 6S, 10CCC
$4_f + 1^5$	4438	2B, 4F, 3D, 5G, 6M, 10FFF
$2^2 + 1^5$	3847	2B, 4F, 3C, 5F, 6Q, 10B
$2 + 2' + 1^5$	3868	2B, 4F, 3C, 5G, 6Q, 10FFF
$6 + 1^3$	1556	2A, 4E, 3C, 5G, 6L, 10TT
$4_f + 2 + 1^3$	1504	2A, 4E, 3C, 5D, 6L, 10YY
$2^3 + 1^3$	951	2A, 4E, 3B, 5E, 6P, 10BBB

$2^2 + 2' + 1^3$	934	2A, 4E, 3B, 5D, 6P, 10YY
$6 + 2 + 1$	2937	2B, 4B, 3B, 5D, 6H, 10MM
$4_f^2 + 1$	3500	2B, 4B, 3C, 5C, 6J, 10FF
$4_f + 2^2 + 1$	2900	2B, 4B, 3B, 5B, 6H, 10O
$4_f + 2 + 2' + 1$	2918	2B, 4B, 3B, 5C, 6H, 10FF
$2^4 + 1$	2305	2B, 4B, 3A, 5A, 6A, 10A
$2^3 + 2' + 1$	2324	2B, 4B, 3A, 5B, 6A, 10O
$2^2 + 2'^2 + 1$	2342	2B, 4B, 3A, 5C, 6A, 10FF
$2^2 + 3 + 1^2$	3088	2B, 4C, 3B, 5D, 6K, 10PP
$2^2 + 3' + 1^2$	3105	2B, 4C, 3B, 5E, 6K, 10XX
$2^2 + 4 + 1$	3052	2B, 4C, 3B, 5B, 6K, 10II
$2^2 + 5$	2476	2B, 4C, 3A, 5B, 6I, 10II
$2 + 3 + 1^4$	1419	2A, 4D, 3C, 5G, 6R, 10DDD
$2 + 3' + 1^4$	1401	2A, 4D, 3C, 5F, 6R, 10EEE
$2 + 4 + 1^3$	1368	2A, 4D, 3C, 5D, 6R, 10AAA
$2 + 5 + 1^2$	786	2A, 4D, 3B, 5D, 6O, 10AAA
$2 + 3^2 + 1$	785	2A, 4D, 3B, 5D, 6O, 10ZZ
$2 + 3 + 3' + 1$	786	2A, 4D, 3B, 5D, 6O, 10AAA
$2 + 3'^2 + 1$	800	2A, 4D, 3B, 5E, 6O, 10WW
$2 + 3 + 4$	764	2A, 4D, 3B, 5C, 6O, 10KK
$2 + 3' + 4$	750	2A, 4D, 3B, 5B, 6O, 10JJ
$4_f + 3 + 1^2$	3665	2B, 4C, 3C, 5D, 6J, 10QQ
$4_f + 4 + 1$	3645	2B, 4C, 3C, 5C, 6J, 10OO
$4_f + 5$	3063	2B, 4C, 3B, 5C, 6H, 10OO
$2 + 2' + 3 + 1^2$	3089	2B, 4C, 3B, 5D, 6K, 10QQ
$2 + 2' + 4 + 1$	3069	2B, 4C, 3B, 5C, 6K, 10OO
$2 + 2' + 5$	2493	2B, 4C, 3A, 5C, 6I, 10OO
$6 + 3$	633	2A, 4A, 3B, 5D, 6F, 10HH
$4_f + 2 + 3$	613	2A, 4A, 3B, 5C, 6F, 10U
$4_f + 2 + 3'$	598	2A, 4A, 3B, 5B, 6F, 10Z
$2^3 + 3$	19	2A, 4A, 3A, 5B, 6C, 10D
$2^3 + 3'$	3	2A, 4A, 3A, 5A, 6C, 10N
$2^2 + 2' + 3$	37	2A, 4A, 3A, 5C, 6C, 10U
$2^2 + 2' + 3'$	22	2A, 4A, 3A, 5B, 6C, 10Z

Remark 4.19 Since Table 4.18 shows that except for representations $2 + 5 + 1^2$ and $2 + 3 + 3' + 1$, every representation affords a different fusion pattern, we see that any two SL(2, 5)-subgroups of $\mathcal{A}$ which do not have fusion pattern 786 (2A, 4D, 3B, 5D, 6O, 10AAA) and are not conjugate in $\mathcal{A}$ are not conjugate in G. Hence Table 4.18 may be regarded as a list of the G-conjugacy classes of SL(2, 5)-subgroups of $\mathcal{A}$ with the possible exception that there may be a G-fusion of the two $\mathcal{A}$-classes of SL(2, 5)-subgroups of $\mathcal{A}$ with fusion pattern 786 (2A, 4D, 3B, 5D, 6O, 10AAA).

We include the definitions from [Gr '91](2.3) of the Hamming codes and the tetracode in Definition 4.20.

Definition 4.20 (Famous codes.) The *Hamming code* (*with parameters* [7, 3, 4]) is the unique self-orthogonal binary code with parameters [7, 3, 4]. Its group is GL(3, 2) and the code may be thought of as the set of complements to linear subspaces of codimension 1 in $\mathbf{F}_2^3$ together with the empty set, with Boolean sum as the addition. The *extended Hamming code* is the unique self-orthogonal [8, 4, 4]-binary code; it is spanned by a copy of the Hamming code supported on seven of the alphabet letters and (1, 1, 1, 1, 1, 1, 1, 1). Its group is AGL(3, 2) $\cong 2^3$: GL(3, 2).

The *tetracode* is the unique self-orthogonal length 4 ternary code; equivalently, it is the unique [4, 2, 3] ternary code. Any nonzero vector has weight 3 and the standard version of this code is spanned by (s, a, a+s, a + 2s), for a, s ε $\mathbf{F}_3$. Its group is GL(2, 3).

Remark 4.21 Let X_i, i= 1, ..., 4, be the four central factors of Δ, ($<z_i>$ = $Z(X_i)$). Now the structure of W_{E_8} implies that $N(\Delta)/C(\Delta) \cong$ GL(2, 3), which acts faithfully on Δ and induces Σ_4 on the set of four factors. In fact, GL(2, 3) is the group of a tetracode via its monomial action based on $\{z_i \mid i = 1,..., 4\}$. Thus, the relations satisfied by $\{z_i \mid i = 1,..., 4\}$ correspond to tetracode words. This is so because the minimum length of a relation is 3 and $rk(Z(C(U))) = 2$. (This remark is lifted almost verbatim from [Gr '91](11.1)).

Remark 4.22 Let η be a 12-dimensional embedding of Alt_5 (resp. SL(2, 5)) into $S := \prod_{i=1}^{4} SL(3, \mathbb{C})$. By Lemmas 4.12 and 4.13, since η affords a single S-class of Alt_5 (resp. SL(2, 5)) subgroups of S, η affords a single Δ-class of Alt_5 (resp. SL(2, 5)) subgroups of Δ. By Remark 4.21, a permutation of the factors of S doesn't change the G-class of $\eta(Alt_5)$ (resp. $\eta(SL(2, 5))$) (and therefore doesn't change the fusion pattern).

Table 4.23 G-conjugacy classes of Alt_5 subgroups of Δ.

The following is a list of all equivalence classes of embeddings (see Notation 4.15) of Alt_5 into $S := \prod_{i=1}^{4} SL(3, \mathbb{C})$ modulo outer twists (see Remark 4.11) and permutations of the factors of S (an outer twist switches the multiplicities of the constituents 3 and 3' which affords a fusion pattern which is in the same equivalence class as the original fusion pattern in the sense of Remark 4.8). By Remarks 4.21 and 4.22, each class corresponds to a unique G-class of Alt_5 subgroups of Δ whose fusion pattern in G is given in the second column. If V_i, $i \in \{1, 2, 3, 4\}$ is a natural 3-dimensional module for the ith factor of Δ, then the fusion pattern is obtained from the formula

$$\chi|_\Delta = \sum_{i=1}^{4} \text{adjoint}(sl_3(V_i)) + \sum_{h \in \mathcal{J}} V_1^{h_1} \otimes V_2^{h_2} \otimes V_3^{h_3} \otimes V_4^{h_4}$$

where h_i is the ith coordinate in h for i = 1, ..., 4, $\mathcal{J}$ is the tetracode, and

$$V_i^{h_i} = \begin{cases} 1 & \text{if } h_i = 0 \\ V_i & \text{if } h_i = 1 \\ V_i^* & \text{if } h_i = -1 \end{cases}.$$

Class of embeddings	Corresponding fusion pattern	
$3 + 1^3 + 1^3 + 1^3$	769	2A, 3D, 5H
$3^2 + 1^3 + 1^3$	1312	2B, 3C, 5F
$3 + 3' + 1^3 + 1^3$	1341	2B, 3C, 5G
$3^3 + 1^3$	302	2A, 3B, 5E
$3^2 + 3' + 1^3$	272	2A, 3B, 5D
3^4	785	2B, 3A, 5A
$3^3 + 3'$	815	2B, 3A, 5B
$3^2 + 3'^2$	844	2B, 3A, 5C

Table 4.24 G-conjugacy classes of SL(2, 5) subgroups of Δ.

The following is a list of all equivalence classes of embeddings (see Notation 4.15) of SL(2, 5) into $S := \prod_{i=1}^{4} SL(3, \mathbb{C})$ modulo outer twists (see Remark 4.11) and permutations of the factors of S (an outer twist switches the multiplicities of the constituents 3 and 3' and 2 and 2' which affords a fusion pattern which is in the same equivalence class as the original fusion pattern in the sense of Remark 4.8). By Remarks 4.21 and 4.22, each class corresponds to a unique G-class of SL(2, 5) subgroups of Δ whose fusion pattern in G is calculated as in Table 4.23, and is given in the second column.

Class of embeddings	Corresponding fusion pattern	
$(2 + 1) + 1^9$	2294	2A, 4G, 3D, 5H, 6S, 10CCC
$(2 + 1)^2 + 1^6$	3847	2B, 4F, 3C, 5F, 6Q, 10B

$(2 + 1) + (2' + 1) + 1^6$	3868	2B, 4F, 3C, 5G, 6Q, 10FFF
$(2 + 1)^3 + 1$	951	2A, 4E, 3B, 5E, 6P, 10BBB
$(2 + 1)^2 + (2' + 1) + 1^3$	934	2A, 4E, 3B, 5D, 6P, 10YY
$(2 + 1)^4$	2305	2B, 4B, 3A, 5A, 6A, 10A
$(2 + 1)^3 + (2' + 1)$	2324	2B, 4B, 3A, 5B, 6A, 10O
$(2 + 1)^2 + (2' + 1)^2$	2342	2B, 4B, 3A, 5C, 6A, 10FF
$(2 + 1) + 3 + 1^6$	1419	2A, 4D, 3C, 5G, 6R, 10DDD
$(2 + 1) + 3' + 1^6$	1401	2A, 4D, 3C, 5F, 6R, 10EEE
$(2 + 1) + 3^2 + 1^3$	785	2A, 4D, 3B, 5D, 6O, 10ZZ
$(2 + 1) + 3 + 3' + 1^3$	786	2A, 4D, 3B, 5D, 6O, 10AAA
$(2 + 1) + 3'^2 + 1^3$	800	2A, 4D, 3B, 5E, 6O, 10WW
$(2 + 1) + 3^3$	170	2A, 4D, 3A, 5B, 6G, 10P
$(2 + 1) + 3^2 + 3'$	188	2A, 4D, 3A, 5C, 6G, 10KK
$(2 + 1) + 3 + 3'^2$	174	2A, 4D, 3A, 5B, 6G, 10JJ
$(2 + 1) + 3'^3$	152	2A, 4D, 3A, 5A, 6G, 10L
$(2 + 1)^2 + 3 + 1^3$	3088	2B, 4C, 3B, 5D, 6K, 10PP
$(2 + 1)^2 + 3' + 1^3$	3105	2B, 4C, 3B, 5E, 6K, 10XX
$(2 + 1)^2 + 3^2$	2491	2B, 4C, 3A, 5C, 6I, 10T
$(2 + 1)^2 + 3 + 3'$	2476	2B, 4C, 3A, 5B, 6I, 10II
$(2 + 1)^2 + 3'^2$	2458	2B, 4C, 3A, 5A, 6I, 10EE
$(2 + 1) + (2' + 1) + 3 + 1^3$	3089	2B, 4C, 3B, 5D, 6K, 10QQ
$(2 + 1) + (2' + 1) + 3^2$	2475	2B, 4C, 3A, 5B, 6I, 10AA
$(2 + 1) + (2' + 1) + 3 + 3'$	2493	2B, 4C, 3A, 5C, 6I, 10OO
$(2 + 1)^3 + 3$	19	2A, 4A, 3A, 5B, 6C, 10D
$(2 + 1)^3 + 3'$	3	2A, 4A, 3A, 5A, 6C, 10N
$(2 + 1)^2 + (2' + 1) + 3$	37	2A, 4A, 3A, 5C, 6C, 10U
$(2 + 1)^2 + (2' + 1) + 3'$	22	2A, 4A, 3A, 5B, 6C, 10Z

Let $\mathcal{H}$ denote the extended Hamming code (see Definition 4.20). Consider the group $R := \prod_{i=1}^{8} SL(2, \mathbb{C})$. For each element $x \in R$, let there correspond an element $a_x \in \mathbb{Z}_2^8$ such that $\pi_i(x) = 1$ if and only if the ith coordinate of a_x is 0, where π_i is the projection map onto the ith factor of R (we are using additive notation for $\mathbb{Z}_2^8$). Let η be a homomorphism from $SL(2, 5)$ into R. Then $\pi_i(\eta)$ is either the trivial representation or one of the representations of SL(2, 5) of degree 2. For each such homomorphism η, let there correspond an element $b_\eta \in Z_2^8$ such that $\pi_i(\eta) = 1$ if and only if the ith coordinate of b_η is 0. Now let v be the projection of R onto Ω. The kernel of v consists of those elements $x \in R$ for which $\pi_i(x) \in Z(\pi_i(R))$ $\forall i$ and the corresponding vector a_x belongs to $\mathcal{H}$ (see [CoGr '87](3.8)). Hence $v\eta$ is a monomorphism from SL(2, 5) into Ω if and only if $b_\eta \notin \mathcal{H}$. If $0 \neq b_\eta \in \mathcal{H}$, then the image of SL(2, 5) under $v\eta$ is isomorphic to Alt_5.

Table 4.25 Fusion patterns of Alt_5 subgroups which occur in Ω.

By the proof of [CoGr '87](3.8), there is a subgroup A of G which acts like Aut($\mathcal{H}$) on the equivalence classes of embeddings (See Notation 4.15) $\eta : SL(2, 5) \to R$. Since A is transitive on the words of $\mathcal{H}$ of weight 4, it suffices to check all equivalence classes of embeddings into $R_1 := \prod_{i=1}^{4} SL(2, \mathbb{C})$ modulo an outer twist (an outer twist switches the multiplicities of the constituents 2 and 2' which affords a fusion pattern in the same equivalence class as the original fusion pattern in the sense of Remark 4.8). Now A is triply transitive on the factors of R, so if $b_\eta = (1,1,1,1,1,1,1,1)$, we may choose to permute any three constituents to the first three factors. Assume the constituents to the first three factors are 2, then there are $\binom{5}{n}$ different permutations of the remaining constituents where n is the multiplicity of the character 2'. Following is a list of all Alt_5 fusion patterns which occur in Ω. Each equivalence class of embeddings $\eta : SL(2, 5) \to R$ listed affords the accompanying fusion pattern. Note that for each such embedding η, $b_\eta \in \mathcal{H}$. If V_i, $i \in \{1, ..., 8\}$ is a natural 2-dimensional module for the ith factor of Ω, then the fusion pattern is obtained from the formula

$$\chi|_\Omega = \sum_{i=1}^{8} \text{adjoint}(sl_2(V_i)) + \sum_{h \in \mathcal{H}} V_1^{h_1} \otimes \ldots \otimes V_8^{h_8}$$

where h_i is the ith coordinate in h for i = 1, ..., 8, $\mathcal{H}$ is the extended Hamming code, and $V_i^{h_i} = \begin{cases} 1 & \text{if } h_i = 0 \\ V_i & \text{if } h_i = 1 \end{cases}$

Class of embeddings	Fusion pattern	
$2^4 + 1^8$	694	2A, 3D, 5H
$2^3 + 2' + 1^8$	753	2A, 3D, 5E
$2^2 + 2'^2 + 1^8$	769	2A, 3D, 5G
2^8	1312	2B, 3C, 5F
$2^7 + 2'$	1177	2B, 3C, 5A
$2^6 + 2'^2$	1252	2B, 3C, 5D
$2^5 + 2'^3$	1207	2B, 3C, 5B
$2^4 + 2'^4$	1341	2B, 3C, 5G
$2^3 + 2' + 2'^3 + 2$	1236	2B, 3C, 5C

Table 4.26 Fusion patterns of SL(2, 5) subgroups which occur in Ω.

As before we have a subgroup A of G which acts like Aut($\mathcal{H}$) on the equivalence classes of embeddings (See Notation 4.15) $\eta : SL(2, 5) \rightarrow R$. Since A is triply transitive on the factors of R, we may choose to permute any three constituents to the first three factors, so when b_η has weight ≤ 3, any permutation of the factors affords the same fusion pattern. For b_η with weight > 3, assume the constituents to the first three factors are 2 (this is always possible (modulo an outer twist) except when b_η has weight 4), then if w := weight of b_η, there are $\binom{8-w}{n}\binom{8-w-n}{m-3}$ permutations of the remaining constituents where n is the multiplicity of the character 2' and m is the multiplicity of the character 2. There are three possibilities for the remaining embeddings of weight 4. Each class of embeddings $\eta: SL(2, 5) \rightarrow R$ listed affords the accompanying fusion pattern. Note that for each embedding η,

$b_\eta \notin \mathcal{H}$. The second column gives the fusion pattern of each class calculated as in Table 4.25.

Class of embeddings	Fusion pattern	
$2 + 1^{14}$	2294	2A, 4G, 3D, 5H, 6S, 10CCC
$2^2 + 1^{12}$	3847	2B, 4F, 3C, 5F, 6Q, 10B
$2 + 2' + 1^{12}$	3868	2B, 4F, 3C, 5G, 6Q, 10FFF
$2^3 + 1^{10}$	951	2A, 4E, 3B, 5E, 6P, 10BBB
$2^2 + 2' + 1^{10}$	934	2A, 4E, 3B, 5D, 6P, 10YY
$2^3 + 1^2 + 2 + 1^6$	2305	2B, 4B, 3A, 5A, 6A, 10A
$2^3 + 1^2 + 2' + 1^6$	2324	2B, 4B, 3A, 5B, 6A, 10O
$2^2 + 2' + 1^2 + 2' + 1^6$	2342	2B, 4B, 3A, 5C, 6A, 10FF
$2^5 + 1^6$	1419	2A, 4D, 3C, 5G, 6R, 10DDD
$2^4 + 2' + 1^6$	1401	2A, 4D, 3C, 5F, 6R, 10EEE
$2^3 + 2' + 2 + 1^6$	1328	2A, 4D, 3C, 5B, 6R, 10P
$2^3 + 2'^2 + 1^6$	1310	2A, 4D, 3C, 5A, 6R, 10L
$2^2 + 2'^2 + 2 + 1^6$	1368	2A, 4D, 3C, 5D, 6R, 10AAA
$2^6 + 1^4$	3088	2B, 4C, 3B, 5D, 6K, 10PP
$2^5 + 2' + 1^4$	3089	2B, 4C, 3B, 5D, 6K, 10QQ
$2^4 + 2'^2 + 1^4$	3105	2B, 4C, 3B, 5E, 6K, 10XX
$2^3 + 2' + 2 + 2' + 1^4$	3052	2B, 4C, 3B, 5B, 6K, 10II
$2^3 + 2'^3 + 1^4$	3141	2B, 4C, 3B, 5G, 6K, 10SS
$2^3 + 1^2 + 2'^3 + 1^2$	3069	2B, 4C, 3B, 5C, 6K, 10OO
$2^7 + 1^2$	19	2A, 4A, 3A, 5B, 6C, 10D
$2^6 + 2' + 1^2$	37	2A, 4A, 3A, 5C, 6C, 10U
$2^5 + 2'^2 + 1^2$	22	2A, 4A, 3A, 5B, 6C, 10Z
$2^4 + 2'^3 + 1^2$	3	2A, 4A, 3A, 5A, 6C, 10N
$2^3 + 2' + 2 + 2'^2 + 1^2$	57	2A, 4A, 3A, 5D, 6C, 10HH

Chapter 5
Fusion patterns of Alt_5 and $SL(2, 5)$ subgroups of H

Notation 5.1 We introduce the notation of [Wood '89]. Let $e_1, \ldots, e_{16}$ be an orthonormal basis for $\mathbb{C}^{16}$. Let C_{16} be the algebra with 1 over $\mathbb{C}$ generated by $\{e_1, \ldots, e_{1\ldots 16}\}$ subject to $e_i e_j + e_j e_i = -2\delta_{ij}$. If I is a subset of $\{1, \ldots, 16\}$, say $I = \{i_1 < \ldots < i_k\}$, then $e_I := e_{i_1} \ldots e_{i_k}$. C_{16} is the Clifford algebra for $\mathbb{C}^{16}$, and Spin(16) is a multiplicative subgroup of the group of units of C_{16}. The center of Spin(16) is equal to $\{\pm 1, \pm e_{1\ldots 16}\} \cong Z_2 \times Z_2$.

Notation 5.2 Let $\tilde{H}$ be the universal cover of $H \cong Hspin(16, \mathbb{C})$, and let $\zeta: \tilde{H} \to H$ be the covering projection. Then $\tilde{H} \cong Spin(16)$ and is also the universal cover of $SO(16, \mathbb{C})$. Let $\pi: \tilde{H} \to SO(16, \mathbb{C})$ be the covering projection from $\tilde{H}$ onto $SO(16, \mathbb{C})$. By [Wood '89], $\ker \pi = \{\pm 1\}$, and so by [CoGr '87](3.3(i)) $\ker \zeta = \{1, e_{1\ldots 16}\}$ or $\{1, -e_{1\ldots 16}\}$. Suppose e is the generator for $\ker \zeta$. Let K be the image of the homomorphism μ with domain $\tilde{H}$ and kernel $\langle -e \rangle$. $K \cong HSpin(16)$ is a subgroup of G and is conjugate in G to H by Lemma 1.24. Let $L \cong Alt_5$, and η a faithful representation of L. Also let $M \cong SL(2, 5)$ and ψ a faithful representation of M. Finally, let z be the central involution of H, z_1 the central involution of S.

Diagram 5.3 The following diagram illustrates the relationships in Notation 5.2.

$$\begin{array}{ccc}
 & H & \\
 & \uparrow \zeta & \\
K \leftarrow & \tilde{H} & \to SO(16, \mathbb{C}) \\
\mu & & \pi
\end{array}$$

Lemma 5.4 Let A_1, A_2 be two finite perfect subgroups of H. Then the following are equivalent:

(a) A_1 is conjugate in H to A_2

(b) $\pi(\zeta^{-1}(A_1))$ is conjugate in $SO(16, \mathbb{C})$ to $\pi(\zeta^{-1}(A_2))$

(c) $\pi(\zeta^{-1}(A_1))'$ is conjugate in $SO(16, \mathbb{C})$ to $\pi(\zeta^{-1}(A_2))'$

Proof. Let $N_1 := \zeta^{-1}(A_1)$, $N_2 := \zeta^{-1}(A_2)$. Then it is clear that A_1 is conjugate to A_2 in H if and only if N_1 is conjugate in $\tilde{H}$ to N_2 and therefore if and only if $\pi(N_1)$ is conjugate to $\pi(N_2)$ in $SO(16, \mathbb{C})$. So we have (a) $\Leftrightarrow$ (b). But then $\pi(N_1)'$ is conjugate in $SO(16, \mathbb{C})$ to $\pi(N_2)'$ and we have (b) $\Rightarrow$ (c).

Let $B_1 := \pi(N_1)'$, $B_2 := \pi(N_2)'$, and suppose that B_1 is conjugate in $SO(16, \mathbb{C})$ to B_2 i.e. suppose (c). Then $\pi^{-1}(B_1)$ is conjugate in $\tilde{H}$ to $\pi^{-1}(B_2)$ and therefore $\pi^{-1}(B_1)'$ is conjugate in $\tilde{H}$ to $\pi^{-1}(B_2)'$. By Lemma 1.10, $B_1 = \pi(N_1)' = \pi(N_1')$ and $B_2 = \pi(N_2)' = \pi(N_2')$ and therefore N_1' is contained in $\pi^{-1}(B_1)$ and N_2' is contained in $\pi^{-1}(B_2)$. By Lemma 1.11, since A_1 and A_2 are perfect, $A_1 = \zeta(N_1')$, and $A_2 = \zeta(N_2')$. If $z \notin A_1$, then $\ker\pi \cap N_1$ is trivial so $\pi|_{N_1}$ is a monomorphism. So $|B_1 = \pi(N_1')| = |N_1'|$ and therefore N_1' has index 2 in $\pi^{-1}(B_1)$. But $\pi^{-1}(B_1)' \subseteq N_1'$ since $\pi^{-1}(B_1)/N_1' \cong \mathbb{Z}_2$ which is abelian. But Lemma 1.11 $\Rightarrow N_1'$ is perfect so $N_1' = \pi^{-1}(B_1)'$. If $z \in A_1$, then N_1' contains $\ker\zeta$, so $|B_1| = (1/2)|N_1'|$ and therefore $|N_1'| = |\pi^{-1}(B_1)|$. So $N_1' = \pi^{-1}(B_1)$. Since N_1' is perfect, so is $\pi^{-1}(B_1)$ so $N_1' = \pi^{-1}(B_1)'$. Similarly, $N_2' = \pi^{-1}(B_2)'$ whether or not $z \in A_2$. But as we said, $\pi^{-1}(B_1)'$ is conjugate in $\tilde{H}$ to $\pi^{-1}(B_2)'$, so N_1 is conjugate in $\tilde{H}$ to N_2'. But then A_1 is conjugate to A_2 in H and hence (c) $\Rightarrow$ (a). ∎

Remark 5.5 Lemma 5.4 implies that classifying Alt_5 and $SL(2, 5)$ subgroups of H up to conjugacy in H reduces to classifying Alt_5 and $SL(2, 5)$ subgroups of $SO(16, \mathbb{C})$. It should be noted however that an Alt_5 (resp. $SL(2, 5)$) subgroup of H may correspond to either an Alt$_5$ or an SL(2, 5) subgroup of SO(16, $\mathbb{C}$), so we must solve both problems in SO(16, $\mathbb{C}$) in order to solve either problem in H.

Lemma 5.6 Every Alt$_5$ subgroup of GL(n, $\mathbb{C}$) is conjugate to an Alt$_5$ subgroup of SO(n, $\mathbb{C}$).

Proof. Since the indicator of any irreducible representation of Alt$_5$ is positive (see [Atlas '85]), any representation of Alt$_5$ is equivalent to a representation over the real field. By [Hall '59](Theorem 16.9.1), every representation of Alt$_5$ is therefore equivalent to an orthogonal representation. Hence, any Alt$_5$ subgroup of GL(n, $\mathbb{C}$) is conjugate to an Alt$_5$ subgroup of O(n, $\mathbb{C}$). Since Alt$_5$ is perfect, any Alt$_5$ subgroup of O(n, $\mathbb{C}$) is a subgroup of SO(n, $\mathbb{C}$).∎

Remark 5.7 By Lemma 5.6, the GL(n, $\mathbb{C}$)-classes of Alt$_5$ subgroups of SO(n, $\mathbb{C}$) are catalogued by the n-dimensional representations of Alt$_5$. However, we are interested in the SO(n, $\mathbb{C}$)-classes of Alt$_5$ subgroups of SO(n, $\mathbb{C}$).

Lemma 5.8 Two finite subgroups of O(n, $\mathbb{C}$) are conjugate in GL(n, $\mathbb{C}$) if and only if they are conjugate in O(n, $\mathbb{C}$).

Proof. See [Tits '55](7.3).∎

Lemma 5.9 Suppose S and T are two conjugate subgroups of O(n, $\mathbb{C}$). If there is an element of O(n, $\mathbb{C}$) of determinant -1 which normalizes T, then S and T are conjugate in SO(n, $\mathbb{C}$).

Proof. Suppose $S^g = T$ where $g \in O(n, \mathbb{C})$. Let $h \in N(T)$ such that $\det(h) = -1$. Then $S^{gh} = T$ and either g or gh has determinant 1. Hence S and T are conjugate in $SO(n, \mathbb{C})$. ∎

Remark 5.10 Any Alt_5 subgroup of $SO(16, \mathbb{C})$ which acts irreducibly on an odd dimensional module is normalized by an element of $O(16, \mathbb{C})$ of determinant -1, namely, the element which is -1 on the odd dimensional module, and 1 on its complement. So, by Lemmas 5.8 and 5.9 and Remark 5.7, the $SO(16, \mathbb{C})$ classes of Alt_5 subgroups which act irreducibly on an odd dimensional module are catalogued by the equivalence classes of 16 dimensional representations of Alt_5 which have odd dimensional constituents. It now remains to determine for such a class of Alt_5 subgroups whether it corresponds to an H-class of Alt_5 subgroups of H or an H-class of $SL(2, 5)$ subgroups of H.

Lemma 5.11 The group $\zeta(\pi^{-1}(\eta(L)))$ contains an Alt_5 subgroup if η has an even number of nontrivial irreducible constituents, and an $SL(2, 5)$ subgroup with central involution z if η has an odd number of nontrivial irreducible constituents.

Proof. Suppose $x \in \eta(L)$ is an element of order 2, and let $u \in \pi^{-1}(x)$. By [St '81], u has order 2 or 4 as the multiplicity of the eigenvalue -1 of x is $\equiv 0$ or 2 (mod 4). Suppose first that u has order 4. Now $\pi^{-1}(\eta(L)) \cong SL(2, 5)$ or $\mathbb{Z}_2 \times \text{Alt}_5$ by Lemma 1.38, and since $\mathbb{Z}_2 \times \text{Alt}_5$ has no element of order 4, $\pi^{-1}(\eta(L)) \cong SL(2, 5)$. Since $\eta(L) \cap \{\pm I_{16}\}$ is trivial, $\pi^{-1}(\eta(L))$ intersects $\ker \zeta$ trivially, and hence $\zeta(\pi^{-1}(\eta(L))) \cong SL(2, 5)$. Since $Z(\pi^{-1}(\eta(L))) = \{\pm 1\}$, $Z(\zeta(\pi^{-1}(\eta(L))))$ contains z.

If u has order 2, then $\pi^{-1}(\eta(L))$ has no elements of order 4, so is isomorphic to $\mathbb{Z}_2 \times \text{Alt}_5$. Again since $\eta(L) \cap \{\pm I_{16}\}$ is trivial, $\pi^{-1}(\eta(L))$

intersects $\ker\zeta$ trivially, and hence $\zeta(\pi^{-1}(\eta(L))) \cong Z_2 \times \text{Alt}_5$ which contains a subgroup isomorphic to Alt_5.

We have shown that $\zeta(\pi^{-1}(\eta(L)))$ contains a subgroup isomorphic to Alt_5 if involutions in $\eta(L)$ have eigenvalue -1 with multiplicity a multiple of 4, and a subgroup isomorphic to $SL(2, 5)$ if involutions in $\eta(L)$ have eigenvalue -1 with multiplicity $\equiv 2 \pmod 4$. Now by Table 4.3, if χ_i is a nontrivial irreducible representation of L, then every involution of $\chi_i(L)$ has eigenvalue -1 with multiplicity 2. Hence, $\zeta(\pi^{-1}(\eta(L)))$ contains a subgroup isomorphic to Alt_5 if η has an even number of nontrivial irreducible constituents, and a subgroup isomorphic to $SL(2, 5)$ if η has an odd number of nontrivial irreducible constituents. ∎

Remark 5.12 There is a 16 dimensional representation of Alt_5, namely the representation 4^4, which has no odd-dimensional constituents, and therefore we cannot say at this point whether or not it corresponds to more than one $SO(16, \mathbb{C})$ class of Alt_5 subgroups. However, all of the $SO(16, \mathbb{C})$ classes of Alt_5 subgroups to which it corresponds are one single class in $GL(16, \mathbb{C})$ and so the elements of each Alt_5 class all have the same eigenvalues. So these Alt_5 subgroups of $SO(16, \mathbb{C})$ correspond to Alt_5 subgroups of H which all have the same fusion pattern in G. In fact, there is only one G-class of Alt_5-subgroups of G with this fusion pattern. This is proved in the treatment of fusion pattern 1341 on page 131. We have included this representation in Table 5.13 below.

Table 5.13 $O(16, \mathbb{C})$-conjugacy classes of embeddings, modulo an outer twist (see Remark 4.11), of $L \cong \text{Alt}_5$ into $SO(16, \mathbb{C})$ which have an even number of nontrivial irreducible constituents.

In this case, $\zeta(\pi^{-1}(\eta(L)))$ contains a subgroup isomorphic to Alt_5 whose fusion pattern in G (= E_8) is calculated as follows and is given in the second column. An outer twist switches the multiplicities of constituents 3 and 3' which affords a fusion pattern which is in the same equivalence class as the original fusion pattern in the sense of Remark 4.8. Let x be an element of H of finite order, then x is conjugate in H to an element of the form $diag(A, A^{-t})$ where $A \in GL(8, \mathbb{C})$. Now x is conjugate to an element of the maximal torus of G and so corresponds to an element a in the Cartan subalgebra of Lie(G) whose action on Lie(G) is as follows: $a.e_r = (a, r)e_r$. So $x.e_r = e^{2\pi i(a,r)}e_r$.

Class of embeddings	Corresponding fusion pattern	
$3^2 + 1^{10}$	769	2A, 3D, 5H
$3 + 3' + 1^{10}$	753	2A, 3D, 5G
$4 + 3 + 1^9$	694	2A, 3D, 5E
$4^2 + 1^8$	753	2A, 3D, 5G
$5 + 3 + 1^8$	302	2A, 3B, 5E
$5 + 4 + 1^7$	361	2A, 3B, 5G
$5^2 + 1^6$	557	2A, 3C, 5G
$3^4 + 1^4$	1312	2B, 3C, 5F
$3^3 + 3' + 1^4$	1252	2B, 3C, 5D
$3^2 + 3'^2 + 1^4$	1341	2B, 3C, 5G
$4 + 3^3 + 1^3$	1177	2B, 3C, 5A
$4 + 3^2 + 3' + 1^3$	1207	2B, 3C, 5B
$5 + 3^3 + 1^2$	785	2B, 3A, 5A
$5 + 3^2 + 3' + 1^2$	815	2B, 3A, 5B
$4^2 + 3^2 + 1^2$	1252	2B, 3C, 5D
$4^2 + 3 + 3' + 1^2$	1236	2B, 3C, 5C
$4^3 + 3 + 1$	1207	2B, 3C, 5B
$5 + 4 + 3^2 + 1$	860	2B, 3A, 5D
$5 + 4 + 3 + 3' + 1$	844	2B, 3A, 5C
4^4	1341	2B, 3C, 5G
$5 + 3 + 4^2$	815	2B, 3A, 5B
$5^2 + 3^2$	1056	2B, 3B, 5D
$5^2 + 3 + 3'$	1040	2B, 3B, 5C

Table 5.14 $O(16, \mathbb{C})$-conjugacy classes of embeddings of L into $SO(16, \mathbb{C})$, modulo an outer twist (see Remark 4.11), which have an odd number of nontrivial irreducible constituents.

In this case, $\zeta(\pi^{-1}(\eta(L)))$ contains a subgroup isomorphic to SL(2, 5) with central involution z whose fusion pattern in G (= E_8) is calculated as in Table 5.13 and is given in the second column. An outer twist switches the multiplicities of constituents 2 and 2' which affords a fusion pattern which is in the same equivalence class as the original fusion pattern in the sense of Remark 4.8.

Class of embeddings	Corresponding fusion pattern
$3 + 1^{13}$	3847 2B, 4F, 3C, 5F, 6Q, 10B
$4 + 1^{12}$	3868 2B, 4F, 3C, 5G, 6Q, 10FFF
$5 + 1^{11}$	4438 2B, 4F, 3D, 5G, 6M, 10FFF
$3^3 + 1^7$	3088 2B, 4C, 3B, 5D, 6K, 10PP
$3^2 + 3' + 1^7$	3105 2B, 4C, 3B, 5E, 6K, 10XX
$4 + 3^2 + 1^6$	3089 2B, 4C, 3B, 5D, 6K, 10QQ
$4 + 3 + 3' + 1^6$	3141 2B, 4C, 3B, 5G, 6K, 10SS
$5 + 3^2 + 1^5$	3665 2B, 4C, 3C, 5D, 6J, 10QQ
$5 + 3 + 3' + 1^5$	3717 2B, 4C, 3C, 5G, 6J, 10SS
$4^2 + 3 + 1^5$	3052 2B, 4C, 3B, 5B, 6K, 10II
$4^3 + 1^4$	3069 2B, 4C, 3B, 5C, 6K, 10OO
$5 + 4 + 3 + 1^4$	3628 2B, 4C, 3C, 5B, 6J, 10II
$5^2 + 3 + 1^3$	2476 2B, 4C, 3A, 5B, 6I, 10II
$5 + 4^2 + 1^3$	3645 2B, 4C, 3C, 5C, 6J, 10OO
$5^2 + 4 + 1^2$	2493 2B, 4C, 3A, 5C, 6I, 10OO
$3^5 + 1$	2491 2B, 4C, 3A, 5C, 6I, 10T
$3^4 + 3' + 1$	2458 2B, 4C, 3A, 5A, 6I, 10EE
$3^3 + 3'^2 + 1$	2476 2B, 4C, 3A, 5B, 6I, 10II
$5^3 + 1$	3063 2B, 4C, 3B, 5C, 6H, 10OO
$4 + 3^4$	2475 2B, 4C, 3A, 5B, 6I, 10AA
$4 + 3^3 + 3'$	2511 2B, 4C, 3A, 5D, 6I, 10MM
$4 + 3^2 + 3'^2$	2493 2B, 4C, 3A, 5C, 6I, 10OO

Next, we classify SL(2, 5) subgroups of $SO(16, \mathbb{C})$.

Lemma 5.15 Let $O(n, \mathbb{C})$ act in the natural way on a nonsingular n-dimensional module V. Let W be a submodule of V, irreducible with respect to the faithful action of $F \leq O(n, \mathbb{C})$ where $F \cong SL(2, 5)$. Then any decomposition of V into a direct sum of modules irreducible with respect to the action of F has at least two summands isomorphic to W.

Proof. We have a nonsingular bilinear form $B: V \times V \to \mathbb{C}$. Consider the restriction $B|_W: W \times W \to \mathbb{C}$. Since W is irreducible, either $W \cap W^\perp = 0$ or $W \subseteq W^\perp$. If $W \not\subseteq W^\perp$ then $B|_W$ is nonsingular, but this is impossible because the indicator of W is negative since the action of F is faithful, and there is therefore no nonsingular symmetric bilinear form on W. So $W \subseteq W^\perp$. Now by [Artin '57] (1.11), $V/W^\perp$ is the dual of W and $V/W^\perp \cong W$. Since V is completely reducible, there exists a subspace U of V such that $V \cong W^\perp \oplus U$ where $U \cong V/W^\perp \cong W$. Since $U \cap W^\perp = 0$, $U \cap W = 0$, and $U \oplus W \cong W \oplus W$ is a submodule of V. ∎

Corollary 5.16 For any embedding ψ of $SL(2, 5)$ into $O(n, \mathbb{C})$, if η is the character afforded by ψ, then (η, η_i) is even where η_i is any irreducible faithful character of $SL(2, 5)$.

Proof. The result is trivially true for $n = 1$. Let $M \cong SL(2, 5)$, V as in Lemma 5.15, and suppose W is a submodule of V irreducible with respect to the faithful action of $\psi(M)$. Then by Lemma 5.15, any decomposition of V into irreducible submodules contains a summand U isomorphic to $W \oplus W$. Consider the quotient module V/U. It is an orthogonal space since U is a direct sum of irreducibles and therefore a direct summand of V. Hence, by induction, any decomposition of V/U into a direct sum of irreducibles has an even number of irreducible modules of each isomorphism type for which the action of $\psi(M)$ is faithful. But then the same is true of $V \cong V/U \oplus W \oplus W$. ∎

CONJUGACY OF Alt$_5$ AND SL(2, 5) SUBGROUPS OF E$_8(\mathbb{C})$

Remark 5.17 Since SL(2, 5) is perfect, any embedding of SL(2, 5) into O(n, $\mathbb{C}$) is an embedding into SO(n, $\mathbb{C}$). By Lemmas 5.8 and 5.9, any embedding of SL(2, 5) into SO(n, $\mathbb{C}$) with odd-dimensional irreducible constituents represents a single SO(16, $\mathbb{C}$)-class of SL(2, 5)-subgroups of SO(16,$\mathbb{C}$).

Lemma 5.18 Let M be as in Notation 5.2. The group $\zeta(\pi^{-1}(\psi(M)))$ contains a subgroup isomorphic to SL(2, 5) if $Z(\psi(M)) \cap \{\pm I_{16}\} = \{I_{16}\}$. If $Z(\psi(M)) = \{\pm I_{16}\}$, then $\zeta(\pi^{-1}(\psi(M)))$ contains an Alt$_5$ subgroup and $\mu(\pi^{-1}(\psi(M)))$ contains an SL(2, 5) subgroup or vice versa. In either case, $\mu(\pi^{-1}(\psi(M)))$ has a conjugate in H.

Proof. Suppose first that $Z(\psi(M)) \cap \{\pm I_{16}\} = \{I_{16}\}$. Then the intersection of $\pi^{-1}(\psi(M))$ with kerζ is trivial, and therefore $\zeta(\pi^{-1}(\psi(M))) \cong Z_2 \times SL(2, 5)$ which contains a subgroup isomorphic to SL(2, 5).

Next, suppose that $Z(\psi(M)) = \{\pm I_{16}\}$. Then $\pi^{-1}(\psi(M)) \cong Z_2 \times SL(2, 5)$ contains kerζ. Let x be an element of $\psi(M)$ of order four. Now $\pi^{-1}(x)$ consists of two elements of order four, both of which have the same square, say, a. Then a is the generator of the center of the SL(2, 5) factor of $\pi^{-1}(\psi(M))$. Now <a> is either kerζ or kerμ. If <a> = kerζ then $\zeta(\pi^{-1}(\psi(M))) \cong Z_2 \times$ Alt$_5$ and $\mu(\pi^{-1}(\psi(M))) \cong SL(2, 5)$. But then $\mu(\pi^{-1}(\psi(M)))$ is conjugate to an SL(2, 5) subgroup of H since K and H are conjugate. The reverse is true if <a> = kerμ. ∎

Table 5.19 O(16, $\mathbb{C}$)-conjugacy classes of embeddings of M $\cong$ SL(2,5) into SO(16, $\mathbb{C}$), modulo an outer twist (see Remark 4.11), which have no nonfaithful constituents.

In this case, either $\zeta(\pi^{-1}(\psi(M)))$ or $\mu(\pi^{-1}(\psi(M)))$ contains a subgroup isomorphic to Alt$_5$ whose fusion pattern in G (= E$_8$) is calculated as in Table

5.13 and is given in the second column. An outer twist switches the multiplicities of constituents 2 and 2' which affords a fusion pattern which is in the same equivalence class as the original fusion pattern in the sense of Remark 4.8. The SL(2, 5) subgroups which correspond to these classes of embeddings are given in Table 5.20.

Class of embeddings	Corresponding fusion pattern	
$6^2 + 2^2$	272	2A, 3B, 5D
4_f^4	557	2A, 3C, 5G
$4_f^2 + 2^4$	302	2A, 3B, 5E
$4_f^2 + 2^2 + 2'^2$	361	2A, 3B, 5G
2^8	769	2A, 3D, 5H
$2^6 + 2'^2$	694	2A, 3D, 5E
$2^4 + 2'^4$	753	2A, 3D, 5G

Table 5.20 $O(16, \mathbb{C})$-conjugacy classes of embeddings of $M \cong SL(2, 5)$ into $SO(16, \mathbb{C})$ modulo an outer twist (see Remark 4.11).

In this case, either $\zeta(\pi^{-1}(\psi(M)))$ or $\mu(\pi^{-1}(\psi(M)))$ contains a subgroup isomorphic to SL(2, 5) whose fusion pattern in G (= E_8) is calculated as in Table 5.13 and is given in the second column. An outer twist switches the multiplicities of constituents 2 and 2' which affords a fusion pattern which is in the same equivalence class as the original fusion pattern in the sense of Remark 4.8. Note that there are some ambiguities as to which conjugacy class some elements belong to. These are represented in the third column by showing both possibilities for each element. For example, in case 1 it is *a priori* unclear whether the elements of order 4 are of type 4D or 4G. We represent this by writing 4_D^G.

Case	Class of embeddings	Corresponding fusion pattern
1	$2^2 + 1^{12}$	2A, 4_D^G, 3D, 5H, 6_N^S, 10_{UU}^{CCC}
2	$4_f^2 + 1^8$	2B, 4_C^F, 3D, 5G, 6M, 10_{SS}^{FFF}

CONJUGACY OF Alt$_5$ AND SL(2, 5) SUBGROUPS OF E$_8(\mathbb{C})$

3	$2^4 + 1^8$	2B, 4_C^F, 3C, 5F, 6_J^Q, 10_{NN}^B
4	$2^2 + 2'^2 + 1^8$	2B, 4_C^F, 3C, 5G, 6_J^Q, 10_{SS}^{FFF}
5	$6^2 + 1^4$	2A, 4_A^E, 3C, 5G, 6L, 10TT
6	$4_f^2 + 2^2 + 1^4$	2A, 4_A^E, 3C, 5D, 6L, 10_{HH}^{YY}
7	$2^6 + 1^4$	2A, 4_A^E, 3B, 5E, 6_F^P, 10_K^{BBB}
8	$2^4 + 2'^2 + 1^4$	2A, 4_A^E, 3B, 5D, 6_F^P, 10_{HH}^{YY}
9	$6^2 + 2^2$	2B, 4B, 3B, 5D, 6H, 10MM
10	4_f^4	2B, 4B, 3C, 5C, 6J, 10FF
11	$4_f^2 + 2^4$	2B, 4B, 3B, 5B, 6H, 10O
12	$4_f^2 + 2^2 + 2'^2$	2B, 4B, 3B, 5C, 6H, 10FF
13	2^8	2B, 4B, 3A, 5A, 6A, 10A
14	$2^6 + 2'^2$	2B, 4B, 3A, 5B, 6A, 10O
15	$2^4 + 2'^4$	2B, 4B, 3A, 5C, 6A, 10FF
16	$2^2 + 3 + 1^9$	2A, 4E, 3B, 5E, 6_P^O, 10_{BBB}^{WW}
17	$2^2 + 3' + 1^9$	2A, 4E, 3B, 5D, 6_P^O, 10_{ZZ}^{YY}
18	$2^2 + 4 + 1^8$	2A, 4E, 3B, 5D, 6_P^O, 10_{AAA}^{YY}
19	$2^2 + 5 + 1^7$	2A, 4E, 3C, 5D, 6_L^R, 10_{AAA}^{YY}
20	$2^2 + 3^2 + 1^6$	2A, 4D, 3C, 5G, 6_L^R, 10_{TT}^{DDD}
21	$2^2 + 3 + 3' + 1^6$	2A, 4D, 3C, 5D, 6_L^R, 10_{AAA}^{YY}
22	$2^2 + 3'^2 + 1^6$	2A, 4D, 3C, 5F, 6_L^R, 10_{RR}^{EEE}
23	$2^2 + 4 + 3 + 1^5$	2A, 4D, 3C, 5B, 6_L^R, 10_P^Z
24	$2^2 + 4 + 3' + 1^5$	2A, 4D, 3C, 5A, 6_L^R, 10_N^L
25	$2^2 + 4^2 + 1^4$	2A, 4D, 3C, 5D, 6_L^R, 10_{HH}^{AAA}
26	$2^2 + 5 + 3 + 1^4$	2A, 4D, 3A, 5B, 6_G^C, 10_P^Z
27	$2^2 + 5 + 3' + 1^4$	2A, 4D, 3A, 5A, 6_G^C, 10_N^L
28	$2^2 + 5 + 4 + 1^3$	2A, 4D, 3A, 5D, 6_G^C, 10_{HH}^{AAA}

29	$2^2 + 3^3 + 1^3$	2A, 4A, 3A, 5B, 6_G^C, 10_D^{JJ}
30	$2^2 + 3^2 + 3' + 1^3$	2A, 4A, 3A, 5B, 6_G^C, 10_P^Z
31	$2^2 + 3 + 3'^2 + 1^3$	2A, 4A, 3A, 5A, 6_G^C, 10_N^L
32	$2^2 + 3'^3 + 1^3$	2A, 4A, 3A, 5C, 6_G^C, 10_{KK}^U
33	$2^2 + 5^2 + 1^2$	2A, 4D, 3B, 5D, 6_F^O, 10_{HH}^{AAA}
34	$2^2 + 4 + 3^2 + 1^2$	2A, 4A, 3A, 5C, 6_G^C, 10_{KK}^U
35	$2^2 + 4 + 3 + 3' + 1^2$	2A, 4A, 3A, 5D, 6_G^C, 10_{HH}^{AAA}
36	$2^2 + 4 + 3'^2 + 1^2$	2A, 4A, 3A, 5B, 6_G^C, 10_{JJ}^Z
37	$2^2 + 5 + 3^2 + 1$	2A, 4A, 3B, 5C, 6_F^O, 10_{KK}^U
38	$2^2 + 5 + 3 + 3' + 1$	2A, 4A, 3B, 5D, 6_F^O, 10_{HH}^{AAA}
39	$2^2 + 5 + 3'^2 + 1$	2A, 4A, 3B, 5B, 6_F^O, 10_{JJ}^Z
40	$2^2 + 4^2 + 3 + 1$	2A, 4A, 3A, 5B, 6_G^C, 10_Z^{BB}
41	$2^2 + 4^2 + 3' + 1$	2A, 4A, 3A, 5, 6_G^C, 10_{DD}^{HH}
42	$2^2 + 4^3$	2A, 4A, 3A, 5_D^A, 6_G^C, 10_X^N
43	$2^2 + 5 + 4 + 3$	2A, 4A, 3B, 5B, 6_F^O, 10_Z^{BB}
44	$2^2 + 5 + 4 + 3'$	2A, 4A, 3B, 5D, 6_F^O, 10_{DD}^{HH}
45	$2^2 + 3^4$	2A, 4D, 3B, 5D, 6_F^O, 10_{HH}^{ZZ}
46	$2^2 + 3^3 + 3'$	2A, 4D, 3B, 5C, 6_F^O, 10_{KK}^U
47	$2^2 + 3^2 + 3'^2$	2A, 4D, 3B, 5D, 6_F^O, 10_{HH}^{AAA}
48	$2^2 + 3 + 3'^3$	2A, 4D, 3B, 5B, 6_F^O, 10_{JJ}^Z
49	$2^2 + 3'^4$	2A, 4D, 3B, 5E, 6_F^O, 10_K^{WW}
50	$4_f^2 + 3 + 1^5$	2B, 4B, 3B, 5B, 6H, 10_O^{AA}
51	$4_f^2 + 4 + 1^4$	2B, 4B, 3B, 5C, 6H, 10_{OO}^{FF}
52	$4_f^2 + 5 + 1^3$	2B, 4B, 3C, 5C, 6J, 10_{OO}^{FF}
53	$4_f^2 + 3^2 + 1^2$	2B, 4C, 3C, 5D, 6J, 10_{MM}^{QQ}
54	$4_f^2 + 3 + 3' + 1^2$	2B, 4C, 3C, 5C, 6J, 10_{OO}^{FF}
55	$4_f^2 + 4 + 3 + 1$	2B, 4C, 3C, 5B, 6J, 10_{II}^{LL}

56	$4_f{}^2 + 4^2$	$2B, 4C, 3C, 5_C^G, 6J, 10_{OO}^{SS}$
57	$4_f{}^2 + 5 + 3$	$2B, 4C, 3A, 5B, 6I, 10_{II}^{LL}$
58	$2^4 + 3 + 1^5$	$2B, 4B, 3A, 5A, 6_A^I, 10_A^{EE}$
59	$2^4 + 3' + 1^5$	$2B, 4B, 3A, 5C, 6_A^I, 10_T^{FF}$
60	$2^4 + 4 + 1^4$	$2B, 4B, 3A, 5B, 6_A^I, 10_{II}^O$
61	$2^4 + 5 + 1^3$	$2B, 4B, 3B, 5B, 6_K^H, 10_{II}^O$
62	$2^4 + 3^2 + 1^2$	$2B, 4C, 3B, 5D, 6_K^H, 10_{MM}^{PP}$
63	$2^4 + 3 + 3' + 1^2$	$2B, 4C, 3B, 5B, 6_K^H, 10_{II}^O$
64	$2^4 + 3'^2 + 1^2$	$2B, 4C, 3B, 5E, 6_K^H, 10_J^{XX}$
65	$2^4 + 4 + 3 + 1$	$2B, 4C, 3B, 5D, 6_K^H, 10_{QQ}^{CC}$
66	$2^4 + 4 + 3' + 1$	$2B, 4C, 3B, 5G, 6_K^H, 10_S^{SS}$
67	$2^4 + 4^2$	$2B, 4C, 3B, 5_B^E, 6_K^H, 10_{AA}^{XX}$
68	$2^4 + 5 + 3$	$2B, 4C, 3C, 5D, 6_J^D, 10_{QQ}^{CC}$
69	$2^4 + 5 + 3'$	$2B, 4C, 3C, 5G, 6_J^D, 10_S^{SS}$
70	$2^2 + 2'^2 + 3 + 1^5$	$2B, 4B, 3A, 5B, 6_I^A, 10_O^{AA}$
71	$2^2 + 2'^2 + 4 + 1^4$	$2B, 4B, 3A, 5C, 6_I^A, 10_{OO}^{FF}$
72	$2^2 + 2'^2 + 5 + 1^3$	$2B, 4B, 3B, 5C, 6_K^H, 10_{OO}^{FF}$
73	$2^2 + 2'^2 + 3^2 + 1^2$	$2B, 4C, 3B, 5D, 6_K^H, 10_{MM}^{QQ}$
74	$2^2 + 2'^2 + 3 + 3' + 1^2$	$2B, 4C, 3B, 5C, 6_K^H, 10_{OO}^{FF}$
75	$2^2 + 2'^2 + 4 + 3 + 1$	$2B, 4C, 3B, 5B, 6_K^H, 10_{II}^{LL}$
76	$2^2 + 2'^2 + 4 + 4$	$2B, 4C, 3B, 5_C^G, 6_K^H, 10_{OO}^{SS}$
77	$2^2 + 2'^2 + 5 + 3$	$2B, 4C, 3C, 5B, 6_J^D, 10_{II}^{LL}$
78	$6^2 + 3 + 1$	$2A, 4D, 3A, 5B, 6G, 10JJ$
79	$6^2 + 4$	$2A, 4D, 3A, 5C, 6G, 10KK$
80	$4_f{}^2 + 2^2 + 3 + 1$	$2A, 4D, 3A, 5B, 6G, 10_P^{BB}$
81	$4_f{}^2 + 2^2 + 3' + 1$	$2A, 4D, 3A, 5D, 6G, 10_X^{AAA}$

82	$4_f2 + 2^2 + 4$	2A, 4D, 3A, 5_D^A, 6G, 10_{DD}^L
83	$2^6 + 3 + 1$	2A, 4D, 3C, 5G, 6_E^R, 10_{DDD}^Q
84	$2^6 + 3' + 1$	2A, 4D, 3C, 5A, 6_E^R, 10_E^L
85	$2^6 + 4$	2A, 4D, 3C, 5_B^F, 6_E^R, 10_R^{EEE}
86	$2^4 + 2'^2 + 3 + 1$	2A, 4D, 3C, 5B, 6_E^R, 10_P^{BB}
87	$2^4 + 2'^2 + 3' + 1$	2A, 4D, 3C, 5D, 6_E^R, 10_X^{AAA}
88	$2^4 + 2'^2 + 4$	2A, 4D, 3C, 5_D^A, 6_E^R, 10_{DD}^L

The following Lemmas are useful in eliminating the ambiguities in Table 5.20.

Lemma 5.21 Let $M \cong SL(2, 5)$ be a subgroup of G which has a central involution of type 2B. Then it is conjugate to an $SL(2, 5)$ subgroup which has z as its central involution.

Proof. Since the central involution of M has type 2B, it is conjugate to z say by $g \in G$. But then M^g is contained in the centralizer of z, i.e. $M^g \leq H$, and the central involution of M^g is z.∎

Lemma 5.22 Let L be a subgroup of G and suppose C(L) contains an involution x of trace -8, then L is conjugate to a subgroup of H.

Proof. Since C(L) contains $x \in 2B$, $L \leq C_G(C) \leq C_G(x)$, and $C_G(x)$ is conjugate to H.∎

Lemma 5.23 Let L be a subgroup of G and suppose C(L) contains an eights-group E which intersects its connected centralizer nontrivially. Then E, and hence C(L) contains an involution of trace -8. In particular, if the rank of C(L) is ≥ 3, L is conjugate to a subgroup of H.

Proof. Let E be the eights-group contained in C(L). Suppose E is 2A-pure. Then by [CoGr '87] (3.9), $C_G(E) \cong F_4(\mathbb{C}) \times E$. So E intersects the connected component of its connected centralizer trivially, and we have a contradiction. Hence, there is no 2A-pure eights-group in T, so T contains an involution of trace -8. ∎

Lemma 5.24 Let $M \cong SL(2, 5)$ be a subgroup of G whose centralizer has rank $\geq$ 2. Then M is conjugate to a subgroup of H.

Proof. If the central involution x of M has type 2B, then by Lemma 5.21, M is conjugate to a subgroup of H. So we may assume x has type 2A.

If C(M) has rank ≥ 3 then M is conjugate to a subgroup of H by Lemma 5.23. So if C(M) has dimension 9, 11, 12 or 13 or dimension ≥ 15, then C(M) has rank ≥ 3 and M is conjugate to a subgroup of H.

If C(M) has dimension 8 or 14, then it has one of the following types: A_2, $A_1A_1T_2$, A_1T_5, T_8, G_2, $B_2A_1T_1$, B_2T_4, $A_2A_1A_1$, $A_2A_1T_3$, A_2T_6, $A_1A_1A_1A_1T_2$, $A_1A_1A_1T_5$. All of these possibilities have rank ≥ 3 except cases A_2 and G_2, so we may assume C(M) has one of those two types. Let E be a fours-group in $C(M)°$. Then $\langle E, x \rangle$ is an eights-group which intersects its connected centralizer nontrivially, and centralizes M, ($x \notin E$ since $|Z(C(M))| = 3$ and 1 in the respective cases) so by Lemma 5.23, $\langle E, x \rangle$ contains an element of type 2B, and M is conjugate to a subgroup of H.

If C(M) has dimension 10, then it has one of the following types: B_2, A_2T_2, $A_1A_1A_1T_1$, $A_1A_1T_4$. All of these possibilities have rank ≥ 3 except case B_2, so we may assume C(M) has type B_2. Let E be a four group in $C(M)°$. If E is 2A-pure then M is conjugate to a subgroup of $\mathcal{E}$. In that case, since E is contained in $C(\mathcal{E})$ which does not contain x, we have an eights-group $\langle E, x \rangle$ which intersects its connected centralizer nontrivially, and centralizes M. By

Lemma 5.23, $<E, x>$ contains an involution of type 2B, and so M is conjugate to a subgroup of H.

If C(M) has dimension 7 or 5, then it has one of the following types: $A_1A_1T_1$, A_1T_4, T_7, A_1T_2, T_5. But each of these has rank ≥ 3 and hence M is conjugate to a subgroup of H.

If C(M) has dimension 6, 4 or 2, then it has one of the following types: A_1A_1, A_1T_3, T_6, A_1T_1, T_4 or T_2. Those which have rank < 3 are the following: A_1A_1, A_1T_1 and T_2. Each of these has a fours-group E which we assume to be 2A-pure. Since $C(E)°$ then has type E_6T_2, M is contained in a subgroup of type E_6. But then $C(M)°$ has a subgroup of type A_2, and no group of type A_1A_1, A_1T_1 or T_2 contains a subgroup of type A_2 so E is not 2A-pure, and M is conjugate to a subgroup of H. ∎

Lemma 5.25 Let $L \cong Alt_5$ be a subgroup of G whose centralizer dimension is not one of $\{0, 1, 3, 8, 14\}$. Then L is conjugate to a subgroup of H.

Proof. If C(L) has rank 1, then it has dimension 1 or 3. If C(L) has rank 2, then it has type T_2, A_1T_1, A_1A_1, A_2, B_2 or G_2. Hence, if dim C(L) is greater than or equal to 15 or is an element of $\{5, 7, 9, 11, 12, 13\}$, then C(L) has rank at least 3 so contains an involution of type 2B.

Suppose C(L) has rank 2 and that the toral fours-group E is 2A-pure. Then $L \leq C(E)$ which has type E_6T_2, that is, L is conjugate to a subgroup of $\mathcal{E}$. But then C(L) contains a subgroup of type A_2, so C(L) does not have type T_2, A_1T_1, A_1A_1 or B_2 and so does not have dimension 2, 4, 6 or 10. Hence if C(L) has type T_2, A_1T_1, A_1A_1 or B_2, then C(L) contains an element of type 2B and so L is conjugate to a subgroup of H. ∎

Lemma 5.26 Every fusion pattern which occurs in Ω, occurs in H.

Proof. This is clear since $\Omega \leq H$. ∎

Lemma 5.27 Every fusion pattern which occurs in $\mathcal{A}$ and corresponds to a 9-dimensional character with trivial constituents occurs in H.

Proof. H contains a subgroup $A \cong SL(8, \mathbb{C})$ which is also a subgroup of $\mathcal{A}$. So any SL(2, 5) subgroup of A is also a subgroup of both H and $\mathcal{A}$. But there is only one $\mathcal{A}$-class of $SL(8, \mathbb{C})$ subgroups in $\mathcal{A}$, so any SL(2, 5) subgroup of $\mathcal{A}$ which corresponds to a class of embeddings into $SL(9, \mathbb{C})$ which has a trivial constituent is conjugate to a subgroup of A and hence of H. ∎

Lemma 5.28 Suppose M is embedded in H and that the embedding has six constituents of degree 2. Then M is conjugate to a subgroup of Ω.

Proof. There is a toral eights-group E in $C(M)^\circ$ which has four elements of type 2A and three elements of type 2B, and <E, z> is a sixteens-group with a 2B-pure eights-subgroup F such that the involutions of <E, z>\F are all of type 2A. The centralizer of <E, z> has dimension 24, and each of the involutions of <E, z> of type 2A centralizes a group of type $A_1 A_1 D_6$ (since it is contained in a fours-group of type AAB [CoGr '87](3.7)), and therefore is in the center of a group of type $A_1 \circ A_1$. Now consider the four 2A-type involutions in <E, z> which correspond to elements of $SO(16, \mathbb{C})$ for which the multiplicity of -1 is 4. They centralize four disjoint subgroups of type $A_1 A_1$ and hence C(<E, z>) contains a group of type A_1^8. Since dim C(<E, z>) = 24 we have that C(<E, z>) has type A_1^8 and therefore M is conjugate to a subgroup of Ω. ∎

Lemma 5.29 Let $SO(n, \mathbb{C})$ act in the natural way on a nonsingular n-dimensional module $V = U \oplus W$ where U and W are isomorphic, self-dual submodules of V, irreducible with respect to the action of $F \leq SO(n, \mathbb{C})$. Then V can be written as a direct sum of isotropic submodules which are isomorphic to U and W.

Proof. We begin with the special case where $V = U \perp W$. Then U and W are nonsingular. Let $\alpha : U \to W$ be a module isomorphism, and let $g : W \times W \to \mathbb{C}$ be a bilinear form defined by $g(\alpha(u), \alpha(u')) := f(u, u')$, where f is the bilinear form for V. Since $f|_U$ is nonsingular, g is nonsingular. Since U and W are absolutely irreducible, and $f|_W$ is nonsingular, there is a $c \in \mathbb{C}^\times$ such that $f|_W = cg$. So $f(\lambda\alpha(u), \lambda\alpha(u')) = cg(\lambda\alpha(u), \lambda\alpha(u')) = \lambda^2 cg(\alpha(u), \alpha(u')) = \lambda^2 cf(u, u')$ by Schur's Lemma. Now if $\beta = c^{-1/2}\alpha$, then β is an isometry, and the irreducible submodules of $U \perp W$ are U, W and the submodules $V_\lambda := \{(x, \lambda\beta(x)) \mid x \in U\}$ for each $\lambda \in \mathbb{C}$. Consider such a submodule V_λ. If $x, y \in U$, then $(x, \lambda\beta(x))$ and $(y, \lambda\beta(y)) \in V_\lambda$, and $f((x, \lambda\beta(x)), (y, \lambda\beta(y))) = f(x, y) + \lambda^2 f(\beta(x), \beta(y)) = f(x, y)(1 + \lambda^2)$. So if $\lambda = \pm i$, where $i = \sqrt{-1}$, V_λ is an isotropic submodule of V, and $V = V_i \oplus V_{-i}$.

Next suppose $V = U \oplus W$ where at least one of U, W (say, U) is nonsingular. Then since U is a module for $SO(n, \mathbb{C})$, so is $U^\perp$, and $V = U \perp U^\perp$ so we are in the first case.

Now if U is singular, it is isotropic since it is irreducible, so if $V = U \oplus W$ where both U and W are singular, then both are isotropic and the pairing is nonsingular since V is nonsingular. ∎

Lemma 5.30 Suppose V is a direct sum of self-dual submodules, irreducible with respect to the action of $F \leq SO(n, \mathbb{C})$, say $V = U_1 \oplus ... \oplus U_t$, and that $U_i \cong U_j$ as submodules for some $i \neq j$. Then there is a one-dimensional torus in $SO(n, \mathbb{C})$ centralizing F.

Proof. Let U be the submodule $U_i \oplus U_j$. If U is nonsingular, then by Lemma 5.29, we may write U as a nonsingular direct sum of isotropic modules. In this case let W := U.

We may assume then that U is singular. Since U_j is an irreducible module, either $U_j \subseteq U_i^\perp$ or $U_j \cap U_i^\perp = 0$. If $U_j \cap U_i^\perp = 0$, then $U = U_i \oplus U_j$ is nonsingular, which contradicts our assumption. Hence, there are two possibilities:

1) $U = U_i \perp U_j$ where one of U_i or U_j is nonsingular and the other is isotropic, and

2) $U = U_i \perp U_j$ where both U_i and U_j are isotropic.

In case (1) we assume without loss of generality that U_i is isotropic, and argue cases (1) and (2) simultaneously. Let W_i be the dual space of U_i in V. Since U_i is isotropic, $U_i \cap W_i = 0$ and since W_i is dual to U_i, we have a nonsingular space $W := U_i \oplus W_i$. Hence by Lemma 5.29, we may write W as a nonsingular direct sum of isotropic irreducible modules.

Now we may write $V = W \perp W^\perp$ since W is nonsingular. Let W_1 and W_2 be isotropic modules such that $W = W_1 \oplus W_2$. Then if η is an element of $GL(n, \mathbb{C})$ which acts as λ on W_1, λ^{-1} on W_2, and trivially on $W^\perp$, we have $((\lambda w_1, \lambda^{-1} w_2, u), (\lambda w_1', \lambda^{-1} w_2', u')) = \lambda \lambda^{-1}(w_1, w_2') + \lambda \lambda^{-1}(w_1', w_2) + (u, u') = ((w_1, w_2, u), (w_1', w_2', u'))$ where $w_1, w_1' \in W_1$, $w_2, w_2' \in W_2$, and $u, u' \in W^\perp$. So η commutes with the action of F on V, and $\eta \in SO(n, \mathbb{C})$. Since λ is arbitrary in $\mathbb{C}$, we have a one-dimensional torus which commutes with the action of F. ∎

Definition 5.31 Let V be as in the previous Lemma. If U is an irreducible submodule of $V = U_1 \oplus ... \oplus U_t$, then we define the *multiplicity* of U in V (denoted mult(U)) to be the number of irreducible submodules U_i in the direct sum decomposition of V which are isomorphic to U. If W is a submodule of V which contains U, then the multiplicity of U in W is denoted by $\text{mult}_W(U)$.

Lemma 5.32 Let V and F be as in Lemma 5.30. Suppose there are s distinct isomorphism types, let m_i be the multiplicity of the ith isomorphism type and let $n_i := \begin{cases} m_i/2 & \text{if } m_i \text{ is even} \\ (m_i-1)/2 & \text{if } m_i \text{ is odd} \end{cases}$, for $i = 1, ..., s$. Then there is a $\sum_{i=1}^{s} n_i$-dimensional torus T in $SO(n, \mathbb{C})$ which centralizes F. Moreover, there is no torus of larger dimension in $SO(n, \mathbb{C})$ which centralizes F.

Proof. We proceed by induction on the dimension of V. If the dimension of V is 1, then the result is vacuously true. Suppose then that the dimension of V is $n > 1$. If $m_i = 1$ for all $i = 1, ..., s$, then again the result is vacuously satisfied. Suppose $m_i > 1$ for some i. Let η be an element of $SO(n, \mathbb{C})$ which commutes with F. If η acts nontrivially on an irreducible submodule U of V, then it acts by multiplication by a scalar, say λ, on U. If U is nonsingular, then $V = U \perp U^{\perp}$ and $(\lambda u, \lambda u') = \lambda^2(u, u') = (u, u')$ if and only if $\lambda^2 = 1$ where u, u' are arbitrary elements of U. Since such a $\lambda = \pm 1$, there is no torus in $SO(n, \mathbb{C})$ which acts nontrivially on U and centralizes F. Now suppose U is isotropic. Then $\text{mult}(U) > 1$ so by Lemma 5.30, there exists a nonsingular submodule W of V such that $W \cong U \oplus U_1$, where U_1 is isotropic. By Lemma 5.30, there exists a one dimensional torus A in $SO(2(\dim U), \mathbb{C})$ which commutes with the action of F on W, namely the torus of elements which act as $\lambda \in \mathbb{C}$ on U and λ^{-1} on U_1. Since we have dealt with all possible nontrivial actions of $SO(2(\dim U), \mathbb{C})$ on U, there is no larger torus in $SO(2(\dim U), \mathbb{C})$ which acts nontrivially on U and centralizes F. Now $V = W \perp W^{\perp}$, and $W^{\perp}$ is an $(n - \dim W)$-dimensional module with $\text{mult}_{W^{\perp}}(U) =$

$\text{mult}_V(U) - 2$ and $\text{mult}_{W^\perp}(X) = \text{mult}_V(X)$ for all submodules $X \neq U$. So by induction, there is a $((\sum_{i=1}^{s} n_i) - 1)$-dimensional torus B in $SO(n - 2(\dim W), \mathbb{C})$ which centralizes F. Moreover, it is maximal in dimension. So we have a $(\sum_{i=1}^{s} n_i)$-dimensional torus, namely $A \oplus B$ in $SO(n, \mathbb{C})$ which centralizes F, but no torus of larger dimension. ∎

Lemma 5.33 Let T be an r-dimensional torus of $SO(16, \mathbb{C})$. Then $\pi^{-1}(T)$ contains an r-dimensional torus, and does not contain an r+1-dimensional torus.

Proof. Clearly $\pi^{-1}(T)$ does not contain an r+1 dimensional torus or else T would have r+1 dimensions. Let A be a maximal torus of $\tilde{H}$. Then $\pi(A)$ is a maximal torus of $SO(16, \mathbb{C})$ and is therefore conjugate say, by g, to B, a maximal torus of $SO(16, \mathbb{C})$ containing T. Let x be a lift of g in $\tilde{H}$. Now $\pi^{-1}(B)$ contains A^x, so contains an r-dimensional torus which maps to T under π, i.e. $\pi^{-1}(T)$ contains an r-dimensional torus. ∎

Lemma 5.34 Suppose L is an Alt_5, $SL(2, 5)$ or Alt_6 subgroup of $SO(16, \mathbb{C})$ and that T is an r-dimensional torus which centralizes L. Then $\pi^{-1}(T)$ commutes with $\pi^{-1}(L)$.

Proof. Let x be an element of L of odd order. Then $\pi^{-1}(x)$ consists of an element x' of order $|x|$ and an element of order $2|x|$. Let $t \in \pi^{-1}(T)$. Then $\pi(t^{-1}x't) = x$ since $\pi(t)$ commutes with x, so $t^{-1}x't \in \pi^{-1}(x)$ and therefore $t^{-1}x't = x'$ since conjugation preserves order. Hence, $\pi^{-1}(T)$ commutes with every element of $\pi^{-1}(L)$ of odd order. There are $1 + 20 + 24 = 45$ elements of Alt_5 with odd order, and $1 + 80 + 144 = 225$ elements of Alt_6 with odd order, that is, more than half of the elements of both Alt_5 and Alt_6 have odd order so the elements of odd order generate the whole group. If $L \cong SL(2, 5)$, then there are $1 + 20 + 24 = 45$ elements of odd order which is more elements than there are in any of the maximal subgroups of $SL(2, 5)$. So the odd-order elements of L generate $\pi^{-1}(L)$ so $\pi^{-1}(T)$ commutes with $\pi^{-1}(L)$. ∎

Lemma 5.35 Suppose L is a subgroup of H, $C_G(L)°$ contains an n-dimensional torus T and that T contains an element x of type 2B. Then $C_H(L^g)°$ contains an n-dimensional torus for some $g \in G$ such that $L^g \subseteq H$.

Proof. Since x has type 2B, $\exists g \in G$ such that $x^g = z$. Hence $T^g \subseteq C(z) = H$, $L^g \subseteq H$ and $T^g \subseteq C_H(L^g)°$. ∎

Table 5.36 This is Table 5.20 (see page 82) with some of the ambiguities removed.

Case	Representation	Corresponding fusion pattern
1	$2^2 + 1^{12}$	2294 2A, 4G, 3D, 5H, 6S, 10CCC
2	$4_f^2 + 1^8$	4438 2B, 4F, 3D, 5G, 6M, 10FFF
3	$2^4 + 1^8$	3847 2B, 4F, 3C, 5F, 6Q, 10B
4	$2^2 + 2'^2 + 1^8$	3868 2B, 4F, 3C, 5G, 6Q, 10FFF
5	$6^2 + 1^4$	1556 2A, 4E, 3C, 5G, 6L, 10TT
6	$4_f^2 + 2^2 + 1^4$	1504 2A, 4E, 3C, 5D, 6L, 10YY
7	$2^6 + 1^4$	951 2A, 4E, 3B, 5E, 6P, 10BBB
8	$2^4 + 2'^2 + 1^4$	934 2A, 4E, 3B, 5D, 6P, 10YY
9	$6^2 + 2^2$	2937 2B, 4B, 3B, 5D, 6H, 10MM
10	4_f^4	3500 2B, 4B, 3C, 5C, 6J, 10FF
11	$4_f^2 + 2^4$	2900 2B, 4B, 3B, 5B, 6H, 10O
12	$4_f^2 + 2^2 + 2'^2$	2918 2B, 4B, 3B, 5C, 6H, 10FF
13	2^8	2305 2B, 4B, 3A, 5A, 6A, 10A
14	$2^6 + 2'^2$	2324 2B, 4B, 3A, 5B, 6A, 10O
15	$2^4 + 2'^4$	2342 2B, 4B, 3A, 5C, 6A, 10FF
16	$2^2 + 3 + 1^9$	951 2A, 4E, 3B, 5E, 6P, 10BBB

17	$2^2 + 3' + 1^9$	934	2A, 4E, 3B, 5D, 6P, 10YY
18	$2^2 + 4 + 1^8$	934	2A, 4E, 3B, 5D, 6P, 10YY
19	$2^2 + 5 + 1^7$	1504	2A, 4E, 3C, 5D, 6L, 10YY
20	$2^2 + 3^2 + 1^6$	1419	2A, 4D, 3C, 5G, 6R, 10DDD
21	$2^2 + 3 + 3' + 1^6$	1368	2A, 4D, 3C, 5D, 6R, 10AAA
22	$2^2 + 3'^2 + 1^6$	1401	2A, 4D, 3C, 5F, 6R, 10EEE
23	$2^2 + 4 + 3 + 1^5$	1328	2A, 4D, 3C, 5B, 6R, 10P
24	$2^2 + 4 + 3' + 1^5$	1310	2A, 4D, 3C, 5A, 6R, 10L
25	$2^2 + 4^2 + 1^4$	1368	2A, 4D, 3C, 5D, 6R, 10AAA
26	$2^2 + 5 + 3 + 1^4$	170	2A, 4D, 3A, 5B, 6G, 10P
27	$2^2 + 5 + 3' + 1^4$	152	2A, 4D, 3A, 5A, 6G, 10L
28	$2^2 + 5 + 4 + 1^3$	210	2A, 4D, 3A, 5D, 6G, 10AAA
29	$2^2 + 3^3 + 1^3$	19	2A, 4A, 3A, 5B, 6C, 10D
30	$2^2 + 3^2 + 3' + 1^3$	22	2A, 4A, 3A, 5B, 6C, 10Z
31	$2^2 + 3 + 3'^2 + 1^3$	3	2A, 4A, 3A, 5A, 6C, 10N
32	$2^2 + 3'^3 + 1^3$	37	2A, 4A, 3A, 5C, 6C, 10U
33	$2^2 + 5^2 + 1^2$	786	2A, 4D, 3B, 5D, 6O, 10AAA
34	$2^2 + 4 + 3^2 + 1^2$	37	2A, 4A, 3A, 5C, 6C, 10U
35	$2^2 + 4 + 3 + 3' + 1^2$	57	2A, 4A, 3A, 5D, 6C, 10HH
36	$2^2 + 4 + 3'^2 + 1^2$	22	2A, 4A, 3A, 5B, 6C, 10Z
37	$2^2 + 5 + 3^2 + 1$	613	2A, 4A, 3B, 5C, 6F, 10U
38	$2^2 + 5 + 3 + 3' + 1$	633	2A, 4A, 3B, 5D, 6F, 10HH
39	$2^2 + 5 + 3'^2 + 1$	598	2A, 4A, 3B, 5B, 6F, 10Z
40	$2^2 + 4^2 + 3 + 1$	22	2A, 4A, 3A, 5B, 6C, 10Z
41	$2^2 + 4^2 + 3' + 1$	57	2A, 4A, 3A, 5D, 6C, 10HH
42	$2^2 + 4^3$	3	2A, 4A, 3A, 5A, 6C, 10N
43	$2^2 + 5 + 4 + 3$	598	2A, 4A, 3B, 5B, 6F, 10Z
44	$2^2 + 5 + 4 + 3'$	633	2A, 4A, 3B, 5D, 6F, 10HH
45	$2^2 + 3^4$	785	2A, 4D, 3B, 5D, 6O, 10ZZ
46	$2^2 + 3^3 + 3'$	764	2A, 4D, 3B, 5C, 6O, 10KK
47	$2^2 + 3^2 + 3'^2$	786	2A, 4D, 3B, 5D, 6O, 10AAA
48	$2^2 + 3 + 3'^3$	750	2A, 4D, 3B, 5B, 6O, 10JJ
49	$2^2 + 3'^4$	800	2A, 4D, 3B, 5E, 6O, 10WW
50	$4_f^2 + 3 + 1^5$	2900	2B, 4B, 3B, 5B, 6H, 10O
51	$4_f^2 + 4 + 1^4$	2918	2B, 4B, 3B, 5C, 6H, 10FF
52	$4_f^2 + 5 + 1^3$	3500	2B, 4B, 3C, 5C, 6J, 10FF
53	$4_f^2 + 3^2 + 1^2$	3665	2B, 4C, 3C, 5D, 6J, 10QQ

54	$4_f2 + 3 + 3' + 1^2$	3645	2B, 4C, 3C, 5C, 6J, 10OO
55	$4_f2 + 4 + 3 + 1$	3628	2B, 4C, 3C, 5B, 6J, 10II
56	$4_f2 + 4^2$	3717 / 3645	2B, 4C, 3C, 5^G_C, 6J, 10^{SS}_{OO}
57	$4_f2 + 5 + 3$	2476	2B, 4C, 3A, 5B, 6I, 10II
58	$2^4 + 3 + 1^5$	2305	2B, 4B, 3A, 5A, 6A, 10A
59	$2^4 + 3' + 1^5$	2342	2B, 4B, 3A, 5C, 6A, 10FF
60	$2^4 + 4 + 1^4$	2324	2B, 4B, 3A, 5B, 6A, 10O
61	$2^4 + 5 + 1^3$	2900	2B, 4B, 3B, 5B, 6H, 10O
62	$2^4 + 3^2 + 1^2$	3088	2B, 4C, 3B, 5D, 6K, 10PP
63	$2^4 + 3 + 3' + 1^2$	3052	2B, 4C, 3B, 5B, 6K, 10II
64	$2^4 + 3'^2 + 1^2$	3105	2B, 4C, 3B, 5E, 6K, 10XX
65	$2^4 + 4 + 3 + 1$	3089	2B, 4C, 3B, 5D, 6K, 10QQ
66	$2^4 + 4 + 3' + 1$	3141	2B, 4C, 3B, 5G, 6K, 10SS
67	$2^4 + 4^2$	3105	2B, 4C, 3B, 5E, 6K, 10XX
68	$2^4 + 5 + 3$	3665	2B, 4C, 3C, 5D, 6J, 10QQ
69	$2^4 + 5 + 3'$	3717	2B, 4C, 3C, 5G, 6J, 10SS
70	$2^2 + 2'^2 + 3 + 1^5$	2324	2B, 4B, 3A, 5B, 6A, 10O
71	$2^2 + 2'^2 + 4 + 1^4$	2342	2B, 4B, 3A, 5C, 6A, 10FF
72	$2^2 + 2'^2 + 5 + 1^3$	2918	2B, 4B, 3B, 5C, 6H, 10FF
73	$2^2 + 2'^2 + 3^2 + 1^2$	3089	2B, 4C, 3B, 5D, 6K, 10QQ
74	$2^2 + 2'^2 + 3 + 3' + 1^2$	3063 / 3069	2B, 4C, 3B, 5C, 6^H_K, 10OO
75	$2^2 + 2'^2 + 4 + 3 + 1$	3052	2B, 4C, 3B, 5B, 6K, 10II
76	$2^2 + 2'^2 + 4^2$	3141 / 3069	2B, 4C, 3B, 5^G_C, 6K 10^{SS}_{OO}
77	$2^2 + 2'^2 + 5 + 3$	3628	2B, 4C, 3C, 5B, 6J, 10II
78	$6^2 + 3 + 1$	174	2A, 4D, 3A, 5B, 6G, 10JJ
79	$6^2 + 4$	188	2A, 4D, 3A, 5C, 6G, 10KK
80	$4_f2 + 2^2 + 3 + 1$	170	2A, 4D, 3A, 5B, 6G, 10P
81	$4_f2 + 2^2 + 3' + 1$	210	2A, 4D, 3A, 5D, 6G, 10AAA
82	$4_f2 + 2^2 + 4$	152	2A, 4D, 3A, 5A, 6G, 10L
83	$2^6 + 3 + 1$	1419	2A, 4D, 3C, 5G, 6R, 10DDD
84	$2^6 + 3' + 1$	1310	2A, 4D, 3C, 5A, 6R, 10L
85	$2^6 + 4$	1401	2A, 4D, 3C, 5F, 6R, 10EEE
86	$2^4 + 2'^2 + 3 + 1$	1328	2A, 4D, 3C, 5B, 6R, 10P
87	$2^4 + 2'^2 + 3' + 1$	1368	2A, 4D, 3C, 5D, 6R, 10AAA
88	$2^4 + 2'^2 + 4$	1310	2A, 4D, 3C, 5A, 6R, 10L

Proof. Cases 1, 35 : The fusion pattern shown is one of the fusion patterns in Ω. By Lemma 5.26, it must also be a fusion pattern in H, but there is no other case listed for which the given fusion pattern is possible. Hence these cases have the fusion patterns shown in the table.

Case 5 : The fusion pattern shown is one of the fusion patterns in $\mathcal{A}$. By Lemma 5.27, it must also be a fusion pattern in H, but there is no other case listed for which the given fusion pattern is possible. Hence this case has the fusion patterns shown in the table.

Cases 2 - 4, 50 - 55, 57 - 73, 75, 77 : These cases are conjugate to SL(2, 5) subgroups with central involution z, and therefore have fusion patterns from either Table 5.14 or Table 5.20 (cases 9-15) with elements of type 4B. In most of these cases, only one of the given possibilities is on one of these tables.

In case 4, there are two possibilities: (2B, 4F, 3C, 5G, 6Q, 10FFF) with connected centralizer of dimension 66, and (2B, 4C, 3C, 5G, 6J, 10SS) with connected centralizer of dimension 10. However, by analyzing the representation, we see that the connected centralizer of an element of this class of SL(2, 5) subgroups of H contains a subgroup of type D_4 and therefore has dimension at least 28 which forces case 4 to be (2B, 4F, 3C, 5G, 6Q, 10FFF).

Cases 6 - 8, 16 - 22, 25, 29 - 34, 36, 37, 39, 45 - 49 : In each of these cases there is a torus in the connected centralizer of M which contains an element of type 3A which forces M to be conjugate to an SL(2, 5) subgroup of $\mathcal{A}$. But then we may search the list of fusion patterns which occur in $\mathcal{A}$ (Table 4.18), and in each of these cases, only one of the possibilities is the fusion pattern of an SL(2, 5) subgroup of $\mathcal{A}$.

Cases 83 - 88: In each of these cases, the embedding has six constituents of degree 2, so, by Lemma 5.28, in each of these cases M is conjugate to a subgroup of Ω. But then by Table 4.26, there is only one possible fusion pattern in each of these cases.

Case 23: Suppose the fusion pattern includes 6R, 10Z. Then C(M) has dimension 17. But by Lemma 5.35, C(M) has rank three so C(M) has type G_2A_1. Now the centralizer of an element x of M of type 10Z has type $A_4A_1A_1T_1$, and contains C(M), but none of the factors of C(x) contains a group of type G_2 so this fusion pattern is impossible.

Suppose the fusion pattern includes 6L and 10Z. Then C(M) has dimension 9, but from the embedding we see that C(M) contains a group of type B_2 which has dimension 10, so this fusion pattern is impossible.

For each of the following cases, there is a fusion pattern listed in Table 5.20 which does not occur in Table 4.9 and therefore there are no SL(2, 5) subgroups of G with those fusion patterns in G: case 23 with 6L, 10P, case 24 with 6L, 10L, case 24 with 6L, 10N.

For those SL(2, 5) subgroups which have an element of type 3D in their centralizer, we can do a similar calculation with the S-constituents of χ. For these calculations, since one G-class may be a union of more than one S-class, each with different trace, we specify the classes in S we are using. The following fusion patterns were then eliminated:

Case 24 with 6R[G], 10N[G]. Since there is only one S-class in the 6R G-class, by Table 1.19, M has elements of type 6GG[S], 2B[S] and 3D[S]. In addition, M has elements of type 4F[S] or 4G[S] and 5A[S]. Now it is not clear whether the trace given in Table 1.20 is the trace of an element of type 10N or its cube so we calculate it both ways. If we assume that we have elements of type 4F[S] and that the trace given in Table 1.20 is the trace of an element of type 10N, then $(\kappa, 3') = -2$, and if we assume it is the trace of the cube of an element of type 10N, then $(\kappa, 2) = -1$. If we assume that we have elements of type 4G[S], then (κ, 1) = -4. So case 24 does not have this fusion pattern, and therefore case 24 has elements of type 6R and 10L. (We eliminated the other two possibilities with (χ, 2')).

Case 28 with 6C, 10HH. Since the elements of M of order 4 have type 4D[G], the elements of M of order 2 have type 2B[S]. Hence, if $g \in M$ has order 10, then $\xi(g) = -3 + \tau$, $17 + 31\tau$ or $17 - 9\tau$ by Table 1.20. In the first case, g^2 has type 5F[S]. In the second and third cases g^2 has type 5E[S]. Moreover, the elements of order six in M have type 6D[S]. Elements of order 3 are in the class 3A[S] and elements of order 4 are either in class 4F[S] or 4G[S]. If we assume we have elements of type 4F then (ξ, 1) = 9, 11 or 7 (depending upon which class of elements of order 10 are represented), but then (ξ, 1) > (χ, 1) (= 8). If (ξ, 1) = 7, then C(M) has rank at least three, but by Lemmas 5.35, 5.32 and 5.34 and Table 5.20, C(M) has rank two. Similarly if we assume we have elements of type 4G[S], then (ξ, 1) = 25, 27 or 23 which, in any case is greater than (χ, 1). So case 28 does not have this fusion pattern.

Case 28 with 6G, 10HH. Everything is the same as in the previous case except that the elements of order 6 are either of type 6J or 6K. If 4F and 6J then (ξ, 1) = 13, 15 or 11 which, in any case, is greater than (χ, 1) (= 4). If 4F and 6K then (ξ, 1) = 7, 9 or 5 which, in any case, is greater than (χ, 1). If 4G and 6J then (ξ, 1) = 29, 31 or 27 which, in any case, is greater than (χ, 1). If 4G and 6K then (ξ, 1) = 23, 25 or 21 which, in any case, is greater than (χ, 1). So case 28 does not have this fusion pattern.

Case 28 with 6C, 10AAA. Since the elements of M of order 4 have type 4D[G], the elements of M of order 2 have type 2B[S]. Hence, if $g \in M$ has order 10, then $\xi(g) = 4 - 3\tau$ or $21 - 7\tau$ by Table 1.20. So the elements of M of order 5 have type 5E[S] in the first case and type 5F[S] in the second. The elements of order 6 have type 6D[S]. If we assume the elements of order 4 have type 4F[S]

then $(\xi, 1) = 5$ or 8. As ξ is the adjoint character for S, if $C_S(M)$ has dimension 5, then it has rank at least three. But by Lemmas 5.32, 5.34 and 5.35 and Table 5.20, $C_G(M)$ has rank two so this fusion pattern is impossible. If $(\xi, 1) = 8$, then by the same argument, $C_S(M)$ has type A_2. But $(\chi, 1) = 10$ so $C_G(M)$ has type B_2. But a group of type A_2 is cannot be a subgroup of a group of type B_2, so this is impossible. If we assume the elements of order 4 have type $4G[S]$, then $(\xi, 1) = 21$ or 24 both of which are greater than $(\chi, 1) (= 10)$. So case 28 does not have this fusion pattern. Hence case 28 has elements of type 6G and 10AAA.

Case 80: Suppose the element $g \in M$ of order 10 has type 10BB. Then by Table 1.20, there are two possibilities for $\kappa(g)$ in S. In either case, M has elements of type $5B[S]$ and $2B[S]$. By Table 1.19, the elements of M of order 4 have type $4F[S]$ or $4G[S]$, the elements of order 6 have type $6J[S]$ or $6K[S]$ and the elements of order 3 have type $3A[S]$. Assume $\kappa(g) = \tau$. If the elements of order 4 are of type $4F[S]$ then $(\kappa, 1) = -4$ or -1 depending on the type of the elements of order 6. If the elements of order 4 are of type $4G[S]$ then $(\kappa, 1) = -12$ or -9 depending on the type of the elements of order 6. Next assume $\kappa(g) = -20 + 11\tau$. If the elements of order 4 are of type $4F[S]$ then $(\kappa, 1) = -7$ or -4 depending on the type of the elements of order 6. If the elements of order 4 are of type $4G[S]$ then $(\kappa, 1) = -15$ or -12 depending on the type of the elements of order 6. All of the calculated inner products are negative and therefore unacceptable. So in case 80 M has elements of type 10P.

Case 81: Suppose the element $g \in M$ of order 10 has type 10X. Then by Table 1.20, $\kappa(g) = 2 - 8\tau$ or $2 - 3\tau$. In the first case, M contains elements of type $5E[S]$ and $2B[S]$. By Table 1.19, the elements of M of order 4 have type $4F[S]$ or $4G[S]$, the elements of order 6 have type $6J[S]$ or $6K[S]$ and the elements of order 3 have type $3A[S]$. If the elements of order 4 have type $4F[S]$, then $(\kappa, 1) = -4$ or -1 depending on the type of the elements of order 6. If the elements of order 4 have type 4G[S], then (κ, 1) = -12 or -9 depending on the type of the elements of order 6. All of the calculated inner products are negative and therefore unacceptable. If $\kappa(g) = 2 - 3\tau$, then M contains elements of type 5F[S] and 2B[S]. The elements of M of orders 4, 3 and 6 are the same as before. If the elements of order 4 have type 4F[S], then (κ, 1) = -6 or -3 depending on the type of the elements of order 6. If the elements of order 4 are of type 4G[S] then (κ, 1) = -14 or -9. All of these calculated inner products are negative also and therefore unacceptable. So in case 81, M has elements of type 10AAA.

Case 82: Suppose the element $g \in M$ of order 10 has type 10DD. Then by Table 1.20, $\kappa(g) = -2$ or $-2 - 5\tau$. In the first case, M contains elements of type 5E[S] and 2B[S]. By Table 1.19, the elements of M of order 4 have type 4F[S] or 4G[S], the elements of order 6 have type 6J[S] or 6K[S] and the elements of order 3 have type 3A[S]. If the elements of order 4 have type 4F[S], then (κ, 1) = -4 or -1 depending on the type of the elements of order 6. If the elements of order 4 have type 4G[S], then (κ, 1) = -12 or -9 depending on the type of the elements of order 6. All of the calculated inner products are negative and therefore unacceptable. If $\kappa(g) = -2 - 5\tau$, then M contains elements of type 5F[S] and 2B[S]. The elements of M of orders 4, 3 and 6 are the same as before. If the elements of order 4 have type 4F[S], then (κ, 1) = -7 or -4 depending on the type of the elements of order 6. If the elements of order 4 have type 4G[S], then (κ, 1) = -15 or -12. All of these calculated inner products are negative also and therefore unacceptable. So in case 82, M has elements of type 10L.

In cases 80, 81 and 82, the fusion pattern is now determined. But for cases 80 and 82, the centralizer has dimension 9, and therefore has rank at least three, but Lemmas 5.32 and 5.34 show that if M corresponds to one of these two classes of embeddings, then $C_H(M)$ has rank two. Now, by Lemma 5.35, there is a G-conjugate of M in H, say M^g such that $C_H(M^g)$ has rank three. Since cases 26 and 27 are the only other cases which could possibly have the same fusion patterns as cases 80 and 82 respectively, the fusion patterns of cases 26 and 27 are also determined.

Suppose M is an SL(2, 5)-subgroup of G and M has elements of type 4A[G]. Then there exists an element $g \in G$ such that if x is an element of M of order four, $x^g \in S$ and has type 4A[S] since 4A[S] $\subseteq$ 4A[G]. Since $(x^g)^2 = z_1$, $M^g \subseteq C_G(z_1)$. But $<x^g> \subseteq S$, so $M^g \subseteq S$. (Reason: Since M^g is quasisimple, the quasiprojection (see Definition 1.28) of M^g into each of the factors is isomorphic to one of: SL(2, 5), $Z_2 \times $ Alt$_5$ (not a possibility for the SL(2, $\mathbb{C}$) factor) or Z_2. Since $<x^g>$ is contained in S, its quasiprojection into the SL(2, $\mathbb{C}$) factor of $C(z_1)$ is Z_2. But that is only possible if the quasiprojection of M^g into the SL(2, $\mathbb{C}$) factor of $C(z_1)$ is Z_2).

Now if g is an element of M of order 10, then $g^5 = z_1$ so the eigenvalues of g are the same as the eigenvalues of g^2 on the module for ξ and on the 3-dimensional fixed point module, and the negatives of the eigenvalues of g^2 on the module for κ. We calculate then the G-trace of all such elements which occur in an SL(2, 5) subgroup with elements of type 4A for which we have not already determined the fusion pattern. For an element of order 10 which squares to an element of type 5A, we get trace $8 + 12\tau$, i.e. the trace of an element of type 10N. For an element of order 10 which squares to an element of type 5B[S], we get trace $11 - 4\tau$, i.e. the trace of an element of type 10Z. For an element of order 10 which squares to an element of type 5C[S], we get trace $87 - 56\tau$, i.e. the trace of an element of type 10D. For an element of order 10 which squares to an element of type 5E[S], we get trace $3 - 3\tau$, i.e. the trace of an element of type 10HH. For an element of order 10 which squares to an element of type 5F[S], we get trace $35 + 33\tau$ which is not the trace of any element of order 10 in G. Any of the unresolved fusion patterns with elements of type 4A therefore contain elements of one of the following: 10N, 10Z, 10D or 10HH.

Also, if M has one of the unresolved fusion patterns with elements of type 4A, then M has elements of type 6F[G] or 6C[G] since these are the only types for which it is possible for the cube of the element to be z_1. Hence case 38 has elements of type 6F and 10HH, case 40 has elements of type 6C and 10Z, case 41 has elements of type 6C and 10HH, case 42 has elements of type 6C, 5A, and 10N, case 43 has elements of type 6F and 10Z and case 44 has elements of type 6F and 10HH. ∎

Remark 5.37 We may now regard Tables 5.13 and 5.19 as a complete list of the fusion patterns of Alt_5-subgroups which occur in H. From these and the other Tables we have constructed, we make a list of the fusion patterns of Alt_5-subgroups which occur in 𝒜, Δ, Ω and H.

Table 5.38 Fusion patterns of Alt_5 subgroups which occur in 𝒜, Δ, Ω, and H.

An x means that an Alt_5 group of the given fusion pattern occurs in the given subgroup of G.

Fusion pattern		𝒜	Δ	Ω	H
272	2A, 3B, 5D	x	x		x
302	2A, 3B, 5E	x	x		x
361	2A, 3B, 5G				x
557	2A, 3C, 5G	x			x
694	2A, 3D, 5E			x	x
753	2A, 3D, 5G	x		x	x
769	2A, 3D, 5H	x	x	x	x
785	2B, 3A, 5A		x		x
815	2B, 3A, 5B		x		x
844	2B, 3A, 5C		x		x
860	2B, 3A, 5D				x
1040	2B, 3B, 5C	x			x

1056	2B, 3B, 5D	x			x
1177	2B, 3C, 5A			x	x
1207	2B, 3C, 5B			x	x
1236	2B, 3C, 5C	x		x	x
1252	2B, 3C, 5D	x		x	x
1312	2B, 3C, 5F	x	x	x	x
1341	2B, 3C, 5G	x	x	x	x

We are now in a position to eliminate all fusion patterns which do not occur in Table 5.38 from consideration.

Notation 5.39 In each case let L be an Alt_5 subgroup of G with the given fusion pattern, Y be a maximal subgroup of L isomorphic to Alt_4, Z be a maximal subgroup of L isomorphic to Dih_{10}, and X be a maximal subgroup of L isomorphic to Dih_6. Also, let $C := C_G(L)$, and T_i denote a torus of rank i.

Case 468 (481): 2A, 3C, 5D

By Table 4.5, C has dimension 17, and therefore by Lemma 5.25, L is conjugate to a subgroup of H. Since this fusion pattern does not appear in our list of fusion patterns in H (Table 5.38), it does not occur in G.

Case 498 (511): 2A, 3C, 5E

By Table 4.5, C has dimension 31, and therefore, by Lemma 5.25, L is conjugate to a subgroup of H. Since this fusion pattern does not appear in our list of fusion patterns in H (Table 5.38), it does not occur in G.

Case 664 (677): 2A, 3D, 5D

By Table 4.5, C has dimension 38, and therefore, by Lemma 5.25, L is conjugate to a subgroup of H. Since this fusion pattern does not appear in our list of fusion patterns in H (Table 5.38), it does not occur in G.

Case 949: 2B, 3A, 5G

By Table 4.5, C has dimension 10. Since, by [CoGr '93] (3.2), $C_G(Y)$ has type A_2A_2, C has rank at most 4. There are three reductive group types of rank ≤ 4 and dimension 10, and hence three possibilities for the type of C: $A_1A_1A_1T_1$, B_2 and A_2T_2. If C has type $A_1A_1A_1T_1$ or A_2T_2, then L is conjugate to a subgroup of H. But then by Lemma 5.35, L has rank 2, so C has type B_2. But a group of type B_2 cannot be a subgroup of a group of type A_2, so this fusion pattern does not occur in G.

Case 981 (994): 2B, 3B, 5A

By Table 4.5, C has dimension 11, and therefore by Lemma 5.25, L is conjugate to a subgroup of H. But by Table 5.38, H has no Alt_5 subgroups with this fusion pattern. Hence G has no Alt_5 subgroups with this fusion pattern.

Case 1011 (1024): 2B, 3B, 5B

By Table 4.5, C has dimension 5, and therefore by Lemma 5.25, L is conjugate to a subgroup of H. But, by Table 5.38, there are no Alt_5 subgroups of H with this fusion pattern. Hence there are no Alt_5 subroups of G with this fusion pattern.

Case 1086 (1099): 2B, 3B, 5E

By Table 4.5, C has dimension 20, and therefore by Lemma 5.25, L is conjugate to a subgroup of H. But by Table 5.38, there are no Alt_5 subgroups of H with this fusion pattern and hence there are no Alt_5 subgroups of G with this fusion pattern.

Case 1145: 2B, 3B, 5G

By Table 4.5, C has dimension 13, and therefore by Lemma 5.25, L is conjugate to a subgroup of H. By Table 5.38 there are no Alt_5 subgroups of H with this fusion pattern, and therefore there are no subgroups of G with this fusion pattern.

Case 1282 (1295): 2B, 3C, 5E

By Table 4.5, C has dimension 23, and therefore by Lemma 5.25, L is conjugate to a subgroup of H. But since this fusion pattern does not occur on Table 5.38, it doesn't occur in G.

Remark 5.40 We have now eliminated from possibility every Alt_5 fusion pattern which does not occur in H. Therefore any Alt_5 fusion pattern which occurs in G occurs in H.

Remark 5.41 We may regard Tables 5.14 and 5.36 as a complete list of the fusion patterns of SL(2, 5)-subgroups which occur in H. From these and the other tables we have constructed, we make a list of the fusion patterns of SL(2, 5)-subgroups which occur in $\mathcal{A}$, Δ, Ω and H.

Table 5.42 Fusion patterns of SL(2, 5) subgroups which occur in $\mathcal{A}$, Δ, Ω and H.

An x means that an SL(2, 5)-group of the given fusion pattern occurs in the given subgroup of G.

Fusion pattern	$\mathcal{A}$	Δ	Ω	H
3 2A, 3A, 4A, 5A, 6C, 10N	x	x	x	x
19 2A, 3A, 4A, 5B, 6C, 10D	x	x	x	x
22 2A, 3A, 4A, 5B, 6C, 10Z	x	x	x	x
37 2A, 3A, 4A, 5C, 6C, 10U	x	x	x	x
57 2A, 3A, 4A, 5D, 6C, 10HH			x	x
152 2A, 3A, 4D, 5A, 6G, 10L		x		x
170 2A, 3A, 4D, 5B, 6G, 10P		x		x
174 2A, 3A, 4D, 5B, 6G, 10JJ		x		x
188 2A, 3A, 4D, 5C, 6G, 10KK		x		x
210 2A, 3A, 4D, 5D, 6G, 10AAA				x
598 2A, 3B, 4A, 5B, 6F, 10Z	x			x
613 2A, 3B, 4A, 5C, 6F, 10U	x			x
633 2A, 3B, 4A, 5D, 6F, 10HH	x			x
750 2A, 3B, 4D, 5B, 6O, 10JJ	x			x
764 2A, 3B, 4D, 5C, 6O, 10KK	x			x
785 2A, 3B, 4D, 5D, 6O, 10ZZ	x	x		x
786 2A, 3B, 4D, 5D, 6O, 10AAA	x	x		x
800 2A, 3B, 4D, 5E, 6O, 10WW	x	x		x
934 2A, 3B, 4E, 5D, 6P, 10YY	x	x	x	x
951 2A, 3B, 4E, 5E, 6P, 10BBB	x	x	x	x
1310 2A, 3C, 4D, 5A, 6R, 10L			x	x
1328 2A, 3C, 4D, 5B, 6R, 10P			x	x
1368 2A, 3C, 4D, 5D, 6R, 10AAA	x		x	x
1401 2A, 3C, 4D, 5F, 6R, 10EEE	x	x	x	x
1419 2A, 3C, 4D, 5G, 6R, 10DDD	x	x	x	x
1504 2A, 3C, 4E, 5D, 6L, 10YY	x			x
1556 2A, 3C, 4E, 5G, 6L, 10TT	x			x
2294 2A, 3D, 4G, 5H, 6S, 10CCC	x	x	x	x
2305 2B, 3A, 4B, 5A, 6A, 10A	x	x	x	x
2324 2B, 3A, 4B, 5B, 6A, 10O	x	x	x	x
2342 2B, 3A, 4B, 5C, 6A, 10FF	x	x	x	x
2458 2B, 3A, 4C, 5A, 6I, 10EE		x		x
2475 2B, 3A, 4C, 5B, 6I, 10AA		x		x
2476 2B, 3A, 4C, 5B, 6I, 10II	x	x		x
2491 2B, 3A, 4C, 5C, 6I, 10T		x		x
2493 2B, 3A, 4C, 5C, 6I, 10OO	x	x		x

2511 2B, 3A, 4C, 5D, 6I, 10MM				x
2900 2B, 3B, 4B, 5B, 6H, 10O	x			x
2918 2B, 3B, 4B, 5C, 6H, 10FF	x			x
2937 2B, 3B, 4B, 5D, 6H, 10MM	x			x
3052 2B, 3B, 4C, 5B, 6K, 10II	x		x	x
3063 2B, 3B, 4C, 5C, 6H, 10OO	x			x
3069 2B, 3B, 4C, 5C, 6K, 10OO	x		x	x
3088 2B, 3B, 4C, 5D, 6K, 10PP	x	x	x	x
3089 2B, 3B, 4C, 5D, 6K, 10QQ	x	x	x	x
3105 2B, 3B, 4C, 5E, 6K, 10XX	x	x	x	x
3141 2B, 3B, 4C, 5G, 6K, 10SS			x	x
3500 2B, 3C, 4B, 5C, 6J, 10FF	x			x
3628 2B, 3C, 4C, 5B, 6J, 10II				x
3645 2B, 3C, 4C, 5C, 6J, 10OO	x			x
3665 2B, 3C, 4C, 5D, 6J, 10QQ	x			x
3717 2B, 3C, 4C, 5G, 6J, 10SS				x
3847 2B, 3C, 4F, 5F, 6Q, 10B	x	x	x	x
3868 2B, 3C, 4F, 5G, 6Q, 10FFF	x	x	x	x
4438 2B, 3D, 4F, 5G, 6M, 10FFF	x			x

We are now in a position to eliminate all fusion patterns which do not occur in Table 5.42 from consideration. By Lemma 5.24, if the dimension of the centralizer of an SL(2, 5) subgroup M of G is two or is greater than four, then M is conjugate to a subgroup of H. Hence we can eliminate all of the fusion patterns from Table 4.9 which do not occur in H from consideration except fusion patterns 28, 30, 44, 599 and 632.

Lemma 5.43 There are no SL(2, 5) subgroups of G with fusion pattern any of 28, 30, 44, 599 or 632.

Proof. Any SL(2, 5) subgroup M of G with one of these fusion patterns has elements of type 4A. Let x be an element of M of type 4A. Then x is conjugate to an element y of S such that $y^2 = z_1$, the central involution of S (and also of M). Hence M^g is contained in $C(z_1) = C(S) \circ S$. But since $y \in M^g$ is in S, $M^g \subseteq S$, but this is not possible as the following analysis shows. In what follows assume that M is an SL(2, 5) subgroup of S.

Case 28: (2A, 3A, 4A, 5B, 6G, 10Z)

Since M has elements of type 6G[G], then by Table 1.19, the elements of order 2 in M have type 2B[S] and the elements of order 4 have type 4B[S]. The elements of order 5 have type 5B[S] or 5C[S], and the elements of order 10 have trace $4 + 2\tau$ or $14 - 8\tau$ by Table 1.20. If the elements of order 6 have type 6J[S] then $(\xi, 1) = 12$ or 18. If the elements of order 6 have type 6K[S] then $(\xi, 1) = 6$ or 12. In all of these cases, $(\xi, 1) > (\chi, 1)$ which is impossible.

Case 30: (2A, 3A, 4A, 5B, 6G, 10JJ)

Since M has elements of type 6G[G], then by Table 1.19, the elements of order 2 in M have type 2B[S] and the elements of order 4 have type 4B[S]. The elements of order 5 have type 5B[S] or 5C[S], and the elements of order 10 have trace $2 + 6\tau$, $2 - 4\tau$ or $22 - 14\tau$ by Table 1.20. If the elements of order 6 have type 6J[S] then $(\xi, 1) = 12$, 16 or 14. If the elements of order 6 have type 6K[S] then $(\xi, 1) = 6$, 10 or 8. In all of these cases, $(\xi, 1) > (\chi, 1)$ which is impossible.

Case 44: (2A, 3A, 4A, 5C, 6G, 10KK)

Since M has elements of type 6G[G], then by Table 1.19, the elements of order 2 in M have type 2B[S] and the elements of order 4 have type 4B[S]. The elements of order 5 have type 5D[S], and the elements of order 10 have trace $1 - 2\tau$, $11 - 2\tau$ or $29 - 18\tau$ by Table 1.20. If the elements of order 6 have type 6J[S] then $(\xi, 1) = 12$, 14 or 16. If the elements of order 6 have type 6K[S] then $(\xi, 1) = 6$, 8 or 10. In all of these cases, $(\xi, 1) > (\chi, 1)$ which is impossible.

Case 599: (2A, 3B, 4A, 5B, 6F, 10BB)

Since M has elements of type 10BB, then by Table 1.20, the elements of M of order 2 have type 2B[S], the elements of order 5 have type 5B[S], and then by Table 1.19, the elements of order 4 have type 4B[S] and the elements of order 6 have type either 6H[S] or 6I[S]. The elements of order 10 have trace 0 or 40 − 20τ by Table 1.20. If the elements of order 6 have type 6H[S], then (ξ, 1) = 12 or 18. If the elements of order 6 have type 6I[S], then (ξ, 1) = 6 or 12. In all of these cases, (ξ, 1) > (χ, 1) which is impossible.

<u>Case 632</u>: (2A, 3B, 4A, 5D, 6F, 10DD)

Since M has elements of type 10DD, then by Table 1.20, the elements of M of order 2 have type 2B[S], the elements of order 5 have type 5E[S] or 5F[S], and then by Table 1.19, the elements of order 4 have type 4B[S] and the elements of order 6 have type either 6H[S] or 6I[S]. The elements of order 10 have trace 3 − τ or 3 + 9τ by Table 1.20. If the elements of order 6 have type 6H[S], then (ξ, 1) = 12 or 18. If the elements of order 6 have type 6I[S], then (ξ, 1) = 6 or 12. In all of these cases, (ξ, 1) > (χ, 1) which is impossible. ∎

Chapter 6
Fusion patterns of Alt$_5$ subgroups of $\mathcal{E}$

Notation 6.1 Let $L \cong \text{Alt}_5$ be a subgroup of $\mathcal{E}$. Let **K** be the 27-dimensional module for $\mathcal{E}$, and ψ its character.

By [CoWa '92], Table 4, $\psi|_L$ is one of the characters listed in Table 6.2.

Table 6.2 The following is a list of the possibilities for $\psi|_L$:

	$\psi\|_L$	Values of $\psi\|_L$ on elements of orders 2, 3 and 5	Classes in $\mathcal{E}$	Values of $\chi\|_L$ on elements of orders 2, 3 and 5	$(\chi\|_L, 1_a)$
a)	$2.1_a + 3_a + 3_b + 4_a + 3.5_a$	3, 0, 2	2A, 3C or 3D, 5E	-2, -3 or 6, 3	1 or 4
b)	$1_a + 3_a + 2.4_a + 3.5_a$	3, 0, -τ	2A, 3C or 3D, 5G	-2, -3 or 6, τ	0 or 3
c)	$1_a + 7.3_a + 5_a$	-5, 0, 8 - 7τ	2B, 3C or 3D, 5I	14, -3 or 6, 22 - 8τ	11 or 14
d)	$3.1_a + 3.3_a + 3.5_a$	3, 0, 6 - 3τ	2A, 3C or 3D, 5F	-2, -3 or 6, 9 - 7τ	2 or 5
e)	$6.1_a + 3.3_a + 3.4_a$	3, 9, 6 - 3τ	2A, 3A, 5F	-2, 15, 9 - 7τ	8
f)	$5.1_a + 3_a + 3_b + 4.4_a$	3, 9, 2	2A, 3A, 5E	-2, 15, 3	7
g)	$3.3_a + 3.3_b + 4_a + 5_a$	-5, 0, 2	2B, 3C or 3D, 5E	14, -3 or 6, 3	5 or 8
h)	$9.1_a + 6.3_a$	3, 9, 9 + 6τ*	2A, 3A, 5A	-2, 15, 16 + 19τ*	16

Here, $\tau = (1 + \sqrt{5})/2$, $\tau^* = (1 - \sqrt{5})/2$

Proof. The columns having to do with trace and conjugacy classes of elements come from Table 1.22. The traces for both ψ and χ are easily computed from the list of eigenvalues and their multiplicities. The names of the conjugacy classes are also from Table 1.22. The final column is obtained by simple calculation. ∎

Remark 6.3 The characters listed in in Table 6.2 do not represent single E-classes of Alt_5-subgroups of E. Rather, each entry in the table represents the set of all Alt_5 subgroups of E which have the same 27-dimensional character. In fact, some characters may correspond to more than one fusion pattern. We will soon see that this is true only for one of the cases in Table 6.2.

We now use Tables 1.22 and 5.38 to determine the G-fusion patterns of the entries in Table 6.2. Since each of the entries except case (a) in Table 6.2 correspond to only one fusion pattern in Table 5.38, there is no ambiguity. In case (a), [CoWa '92] documents the existence of an Alt_5 subgroup with fusion pattern 557 (2A, 3C, 5G) (since (a) is realized in E via a representation of Alt_5 into SL(6, $\mathbb{C}$) which, by Lemma 1.26 is conjugate to a subgroup of A), but there is also the possibility of a subgroup with fusion pattern 361 (2A, 3B, 5G). In fact, if L is an Alt_5 subgroup in H corresponding to the $5 + 4 + 1^7$ class of embeddings, then C(L) has type B_3. But then L is contained in C(C(L)) of type B_4 containing a conjugate of D. Hence L is conjugate to a subgroup of E and E has a subgroup with fusion pattern 361.

Table 6.4 A list of all possible fusion patterns of Alt_5 subgroups which correspond to each of the characters listed in Table 6.2.

Case in Table 6.2	Fusion pattern in E	Fusion pattern in G		$(\chi \mid_L 1)$
(a)	2A, 3C or 3D, 5E	361	2A, 3B, 5G	1
		557	2A, 3C, 5G	4
(b)	2A, 3C, 5G	272	2A, 3B, 5D	0
(c)	2B, 3D, 5I	1312	2B, 3C, 5F	14

(d)	2A, 3C, 5F	302	2A, 3B, 5E	2
(e)	2A, 3A, 5F	694	2A, 3D, 5E	8
(f)	2A, 3A, 5E	753	2A, 3D, 5G	7
(g)	2B, 3D, 5E	1341	2B, 3C, 5G	8
(h)	2A, 3A, 5A	769	2A, 3D, 5H	16

Proof. In each case where more than one fusion pattern is possible, the pattern not listed is not in Table 5.38 and therefore is not in G by Remark 5.40. ∎

Lemma 6.5 Suppose that L is an Alt_5 subgroup of G such that $C(L)°$ has type A_1. Then L is conjugate to a subgroup of $\mathcal{A}$ or H, or L is diagonally embedded into a subgroup W of G of type A_2E_6 such that the 1-dimensional torus in $C(L)°$ is contained in the E_6 factor of W, but $C(L)°$ is not contained in the E_6 factor of W. Moreover, the latter case is unique up to conjugacy.

Proof. Consider the element y of order three in $C(L)°$. If y has type 3C, then C(y) has type D_7T_1 and L is conjugate to a subgroup of H. Likewise if y has type 3A, then C(y) has type A_8 and L is conjugate to a subgroup of $\mathcal{A}$. If y has type 3D, then C(y) has type E_7T_1 and L is conjugate to a subgroup of $\mathcal{S}$, but the orthogonality relations for the character κ yield negative coefficients for the trivial character for all possible fusion patterns of Alt_5 subgroups with centralizer of type A_1. Hence none of these Alt_5 subgroups is conjugate to a subgroup of $\mathcal{S}$. Hence we may assume that y has type 3B and C(y) has type A_2E_6.

Since y is contained in the 1-dimensional torus T of $C(L)°$ and therefore centralizes T, $T \leq C(y)$ of type A_2E_6. Let π_1 be a projection map from L into the A_2 factor X of C(y), π_2 the corresponding projection map from L into the E_6 factor Y of C(y). Let $t \in T, x \in L$. Then $t = t_1t_2$, and $x = x_1x_2$ where

$t_i := \pi_i(t)$ and $x_i := \pi_i(x)$ for $i \in \{1, 2\}$. Now $tx = xt$, that is, $t_1 t_2 x_1 x_2 = t_1 x_1 t_2 x_2 = (x_1^{t_1})(x_2^{t_2}) t_1 t_2 = x_1 x_2 t_1 t_2$, that is, $(x_1^{t_1})(x_2^{t_2}) = x_1 x_2$. So $x_1^{t_1} = x_1 y^i$ and $x_2^{t_2} = x_2 y^j$ for some $i, j \in \{0, 1, 2\}$. But then $i, j = 0$ for $|x_1|$ and $|x_2| = 1, 2, 5$ and therefore for $|x_1|$ and $|x_2| = 3$. So each element of T centralizes $\pi_1(L)$ and $\pi_2(L)$. Now, since X is conjugate to a subgroup of $\mathcal{A}$, $\pi_1(L)$ has fusion pattern 769 (2A, 3D, 5H) by Table 4.16, and therefore has connected centralizer of type E_6, that is, $C(\pi_1(L)) = Y$. But $\pi_2(L) \leq C(\pi_1(L)) = Y$ and therefore is conjugate to a subgroup of $\mathcal{E}$. Moreover, $T \subseteq C(\pi_1(L)) \cap C(\pi_2(L)) = Y \cap C(\pi_2(L)) = C_Y(\pi_2(L))$, and so $\pi_2(L)$ is conjugate to an Alt_5 subgroup of $\mathcal{E}$ with $\mathcal{E}$-centralizer of dimension 1 or 3. But by Table 6.4, there is no Alt_5 subgroup of $\mathcal{E}$ with $\mathcal{E}$-centralizer of dimension 3, so $\pi_2(L)$ has fusion pattern 361 (2A, 3B, 5G).

Now we have $C := C_G(\pi_2(L))^\circ$ has type B_3 and $(C_G(\pi_2(L))^\circ \cap \mathcal{E})^\circ = C_{\mathcal{E}}(\pi_2(L))^\circ$ has type T_1. Now C contains a toral eights-group E, as well as the centralizer of $\mathcal{E}$. Since the centralizer of $\mathcal{E}$ contains a 2A-pure fours-group which is contained in E, E has at least three type 2A involutions. Hence E has either 6, 4 or 3 elements of type 2A and $C_G(E)^\circ$ has type $D_5 T_3$, $A_5 A_1$ or B_4 in the respective cases. Now C is contained in a conjugate of H and its connected centralizer in H has type B_4 so the centralizer of C in G contains a group of type B_4, and therefore $C_G(E)$ contains a group of type B_4 so $C_G(E)^\circ$ does not have type $A_5 A_1$. Now the centralizer in $\mathcal{E}$ of the element of order 2 in $C_{\mathcal{E}}(\pi_2(L))^\circ$ therefore has type $D_5 T_1$ so has type $2B[\mathcal{E}] \subseteq 2B[G]$. Hence $\pi_2(L)$ is a subgroup of the D_5 factor X of the centralizer of the element of order 2 in $C_{\mathcal{E}}(\pi_2(L))^\circ$ and, by Lemma 1.27, X is unique up to $\mathcal{E}$-conjugacy. Since, by Table 5.13, there is only one class of embeddings of Alt_5-subgroups with this fusion pattern in a group of type $4D_5$, this subgroup L is unique up to conjugacy in G. ∎

Chapter 7
Conjugacy classes of Alt_5 subgroups of G

We now consider each of the remaining fusion patterns on a case by case basis. We know, for each fusion pattern, the dimension of the connected centralizer of an Alt_5 subgroup with the given fusion pattern, and we also know the type of the connected centralizers of the maximal subgroups. Clearly, the connected centralizer of an Alt_5 subgroup is a subgroup of each of the connected centralizers of its maximal subgroups. By [CoGr '87](2.3) we know that the connected centralizer of a finite subgroup is reductive, and therefore we can limit our search for the connected centralizer of an Alt_5 subgroup to groups which are reductive subgroups of each of the connected centralizers of the maximal subgroups and have the dimension given in Table 4.5.

Lemma 7.1 Let D be a subgroup of $\mathcal{E}$ of type $2^2 D_4$. Then the center of D has type BBB.

Proof. Suppose $Z(D)$ has type AAA. Then, by [CoGr '87](3.7), $C(Z(D)) = XT$ where X has type E_6, T is a 2-dimensional torus and $Z(D)$ T since $|Z(X)| = 3$. But $D \leq C(Z(D))$, so $D \leq X$. Since $|X \cap T| = 3$, this is impossible.

Next, suppose $Z(D)$ has type AAB. Then, by [CoGr '87](3.7), $C(Z(D)) = X_1 X_2 Y$ where X_1 and X_2 have type A_1 and Y has type D_6, and $Z(D) = Z(Y)$. If $D \leq \mathcal{E}$, then since $Z(D) \leq \mathcal{E}$, $C(\mathcal{E})$ of type $A_2 \subseteq C(Z(D))$ of type $A_1 A_1 D_6$. Therefore $C(\mathcal{E})$ is contained in the D_6 factor of $C(Z(D))$. But then we have a group of type $A_2 D_4$ as a subgroup of a group of type D_6 which is impossible since the determinant of a lattice of type D_6 is 4 while the determinant of a lattice of type A_2D_4 is 12. But by [Gr unp.](8.12), one lattice is a sublattice of another only if the quotient of their determinants is a perfect square. Hence $D \not\leq E$.

Finally, suppose $Z(D)$ has type ABB. Then, by [CoGr '87](3.7), $C(Z(D)) = XT$ where X has type A_7 and T is a 1-dimensional torus. But $Z(X) \cong \mathbb{Z}_8$, and so $Z(D)$ is partially contained in T contradicting the fact that D is contained in the A_7 factor of $C(Z(D))$. ∎

Corollary 7.2 The group $\mathcal{F}$ contains a conjugate of $\mathcal{D}$.

Proof. By [Carter '89](Sect. 3.6), there exists a subgroup D of $\mathcal{F}$ generated by root groups with root lattice M of type D_4. Let L be the root lattice of $\mathcal{F}$. Suppose M is not a direct summand of L. Then $M \leq M_1 \leq L$ where M_1 is a direct summand of L with the same rank as M, and $[M_1 : M] = d \in \mathbb{Z}$. Now det $M_1 = (1/d^2)(\det M)$ and det $M_1 \in \mathbb{Z}$. But det $M = 4$ so det $M_1 = 1$ and M_1 is an even unimodular lattice so dim $M_1 \equiv 0 \pmod 8$, a contradiction since dim $L = 4$. Hence M is a direct summand of L and so by [Gr '91](2.13)(iv), D is simply connected. By Lemma 7.1, $Z(D)$ has type BBB, so by Lemma 1.29, D is conjugate to $\mathcal{D}$. ∎

Notation 7.3 In each case let L have the given fusion pattern, Y be a maximal subgroup of L isomorphic to Alt_4, Z be a maximal subgroup of L isomorphic to Dih_{10}, and X be a maximal subgroup of L isomorphic to Dih_6. Also let $C = C_G(L)$, and T_i denote a torus of rank i. Finally, we call an Alt_5 subgroup of G with 0-dimensional centralizer a *ZDC*.

<u>Case 272 (285)</u>: 2A, 3B, 5D

By Table 4.5, C has dimension 14. Since, by Table 3.11, $C_G(Z)$ has type B_1B_4, C has rank at most 5. Also note that, by Lemma, 1.36, $C_G(Y)$ has type

D_4T_2. There are three reductive group types of rank ≤ 5 and dimension 14, and hence three possibilities for the type of C: G_2, $B_2A_1T_1$ and $A_2A_1A_1$. If C doesn't have type G_2, then rank C 3 and so by Lemma 5.23, L is conjugate to a subgroup of H. But then by Lemma 5.30, there exists a conjugate of L in H with a 4-dimensional torus in its H-centralizer. But by Tables 5.13 and 5.19, there isn't any such subgroup. Hence C has type G_2.

Case 302 (315): 2A, 3B, 5E

By Table 4.5, C has dimension 28 and therefore, by Lemma 5.25, L is conjugate to a subgroup of H. Since C has rank at least 3, then by Lemma 5.35 and Tables 5.13 and 5.19, C has rank 3 or 4. There are two reductive group types of rank 3 or 4 and dimension 28, and hence two possibilities for the type of C:

1) C has type G_2G_2. Now $C \leq C_G(Y)$ of type D_4T_2 by Lemma 1.36 and the semisimple part of C is contained in the semisimple part of $C_G(Y)$, but a group of type G_2G_2 cannot be contained in a group of type D_4 so C does not have this type.

2) C has type D_4. If C has this type, then C is conjugate to $\mathcal{D}$ since the D_4 factor of $C_G(Y)$ is conjugate to $\mathcal{D}$. Therefore if L has this fusion pattern, L is conjugate to a subgroup of $C(\mathcal{D})$ and hence conjugate to a subgroup of $\mathcal{D}$. But there is only one class of embeddings of Alt_5 subgroups of $\mathcal{D}$ with this fusion pattern by Table 5.13, hence only one G-class with this fusion pattern in G.

Case 361: 2A, 3B, 5G

By Table 4.5, C has dimension 21, and therefore by Lemma 5.25, L is conjugate to a subgroup of H. Since C has rank at least 3, then by Lemma 5.35 and Tables 5.13 and 5.19, C has rank 3 so has type B_3 or C_3. By Lemma 1.36,

$C_G(Y)$ has type D_4T_2. Since a group of type D_4 does not contain a group of type C_3, C has type B_3. Now $C \leq C_G(Y)$ is contained in a conjugate of $\mathcal{D}$ since the D_4 factor of $C(Y)$ is conjugate to $\mathcal{D}$, so the connected centralizer of C is a group of type B_4 which contains a conjugate of $\mathcal{D}$. Now there is only one class of embeddings of Alt_5 subgroups with this fusion pattern in a group of type B_4 in H by Table 5.13, and hence only one G-class with this fusion pattern in G.

Case 557: 2A, 3C, 5G

By Table 4.5, C has dimension 24 and therefore by Lemma 5.25, L is conjugate to a subgroup of H. Hence, by Lemma 5.35 and Tables 5.13 and 5.19, C has rank at most 4. Therefore C has one of the following types: B_3A_1, C_3A_1, G_2B_2, or A_4.

1) If C has type B_3A_1, then the B_3 component of C embeds into the A_5 component of $C_G(Y) \cong A_5T_1$ by Lemma 1.36. But a group of type B_3 acts on a 7-dimensional module while a group of type A_5 acts on a 6-dimensional module, so C doesn't have this type.

2) If C has type C_3A_1, then C embeds into the A_5 component of $C_G(Y)$. But a group of type C_3A_1 acts on a 9-dimensional module while a group of type A_5 acts on a 6-dimensional module, so C doesn't have this type.

3) C has type G_2B_2. Then the G_2 component of C embeds into the A_5 component of $C_G(Y)$. But a group of type G_2 acts on a 7-dimensional module while a group of type A_5 acts on a 6-dimensional module, so C doesn't have this type.

4) C has type A_4. By Tables 5.13 and 5.19, there are two $GL(16, \mathbb{C})$-classes of embeddings which correspond to this fusion pattern, namely, $5^2 + 1^6$ and 4_f^4. (Now $5^2 + 1^6$ is actually an $SO(16, \mathbb{C})$-class of embeddings by Remark 5.10, but at this point, we are unable to say the same for 4_f^4). Suppose L

corresponds to the $4f^4$ class of embeddings. Then by Lemmas 5.32 and 5.34, there is a two-dimensional torus in H which centralizes L. Now the connected centralizer in G of L contains a four-dimensional torus and therefore an element of type 2B, but then by Lemma 5.35, $\exists g \in G$ such that $L^g \subseteq H$ and $C_H(L^g)$ contains a four dimensional torus. But then L^g corresponds to the $5^2 + 1^6$ class of embeddings, so there is only one G-class of Alt_5 subgroups with this fusion pattern in G.

<u>Case 694 (707)</u>: 2A, 3D, 5E

By Table 4.5, C has dimension 52, and therefore L is conjugate to a subgroup of H by Lemma 5.25. Since, by Table 3.11, $C_G(Z)$ has type F_4T_1, C has rank at most 5. There is only one reductive group type of rank ≤ 5 and dimension 52 and hence there is only one possibility for the type of C:

1) C has type F_4. Since C is contained in the centralizer of an element of L of type 5E, C is conjugate to $\mathcal{F}$. Hence L is conjugate to a subgroup of $\mathcal{G}$. But, $\mathcal{G}$ is a subgroup of $\mathcal{D}$, and hence L is conjugate to a subgroup of $\mathcal{D}$. By Table 5.13, there is a unique class of embeddings of Alt_5 subgroups with this fusion pattern in $\mathcal{D}$, and hence a unique conjugacy class of subgroups with this fusion pattern in G. ∎

<u>Case 753</u>: 2A, 3D, 5G

By Table 4.5, C has dimension 45, and therefore, by Lemma 5.25, L is conjugate to a subgroup of H. Since, by Table 3.11, $C_G(Z)$ has type D_5T_1, C has rank at most 5. There is only one reductive group type of rank ≤ 5 and dimension 45, and hence only one possibility for the type of C:

1) C has type D_5. By Tables 5.13 and 5.19, there are three $GL(16, \mathbb{C})$-classes of embeddings which correspond to this fusion pattern, namely, $3 + 3'$

$+ 1^{10}$, $4^2 + 1^8$ and $2^4 + 2'^4$. (Now $3 + 3' + 1^{10}$ and $4^2 + 1^8$ are actually SO(16, $\mathbb{C}$)-classes of embeddings by Remark 5.10, but at this point we are unable to say the same for $2^4 + 2'^4$). Suppose L corresponds to the $3 + 3' + 1^{10}$ class of embeddings. Then by Lemmas 5.32 and 5.34, there is a five-dimensional torus in H which centralizes L. Since this torus contains an element of type 3A, L is conjugate to a subgroup of $\mathcal{A}$. Similarly, if L corresponds to the $4^2 + 1^8$ class of embeddings, L is conjugate to a subgroup of $\mathcal{A}$. Since there is only one class of Alt_5 subgroups of $\mathcal{A}$ with this fusion pattern, these two classes of embeddings of L fuse into one G-class in G. Now suppose L corresponds to the $2^4 + 2'^4$ class of embeddings. Then by Lemmas 5.32 and 5.34, there is a four-dimensional torus in H which centralizes L. Now the connected centralizer in G of L contains a five-dimensional torus and therefore an element of type 2B, but then by Lemma 5.35, $\exists g \, \varepsilon \, G$ such that $L^g \subseteq H$ and $C_H(L^g)$ contains a five-dimensional torus. But then L^g corresponds to the class of embeddings corresponding to one of $3 + 3' + 1^{10}$ or $4^2 + 1^8$. Since these two classes fuse in H, there is only one G-class of Alt_5 subgroups in G with this fusion pattern.

<u>Case 769 (782)</u>: 2A, 3D, 5H

By Table 4.5, C has dimension 78 and therefore L is conjugate to a subgroup of H by Lemma 5.25. Since, by Table 3.11, $C_G(Z)$ has type E_6, C has rank at most 5. There are two reductive group types of rank ≤ 5 and dimension 78 and hence there are two possibilities for C:

1) C has type B_6. Then $C_G(Z)$ contains a group of type B_6 and therefore they share a maximal torus. This forces the root lattice for B_6 to be a sublattice of the root lattice for E_6 which is a contradiction since all the roots of E_6 have the same length.

2) C has type E_6. By Tables 5.13 and 5.19, there are two GL(16, $\mathbb{C}$) classes of embeddings which correspond to this fusion pattern, namely, $3^2 + 1^{10}$ and 2^8. (Now $3^2 + 1^{10}$ is actually an SO(16, $\mathbb{C}$)-class of embeddings by Remark 5.10, but at this point, we are unable to say the same for 2^8). Suppose L corresponds to the 2^8 class of embeddings. Then by Lemmas 5.32 and 5.34, there is a four-dimensional torus in H which centralizes L. Now the connected centralizer in G of L contains a six-dimensional torus and therefore an element of type 2B, but then by Lemma 5.35, $\exists g \in G$ such that $L^g \subseteq H$ and $C_H(L^g)$ contains a six-dimensional torus. But then L^g corresponds to the $3^2 + 1^{10}$ class of embeddings, so there is only one G-class of Alt_5 subgroups with this fusion pattern in G.

Lemma 7.4 Suppose C contains an Alt_5 subgroup. Then one of the following things occurs:

    1) L is conjugate to a subgroup of $\mathcal{E}$, and C contains an A_2 subgroup.

    2) L is conjugate to a subgroup of $\mathcal{F}$, and C contains a G_2 subgroup,

    3) C contains a three dimensional torus or

    4) L is conjugate to a subgroup of H.

Proof. Let L' be an Alt_5-subgroup of C. By Remark 5.40, L' has one of the following fusion patterns in G: 272, 302, 361, 557, 694, 753, 769, 785, 815, 844, 860, 1040, 1056, 1177, 1207, 1236, 1252, 1312, 1341. If L' contains an involution of trace -8, then C(L') ($\supseteq$ L) is conjugate to a subgroup of H. The remaining cases are those for which the involutions of L' have type 2A.

Case L' has fusion pattern 272: Then C(L') has type G_2 and is conjugate to $\mathcal{G}$. Therefore since $\mathcal{G}$ is conjugate to a subgroup of $\mathcal{F}$, L is conjugate to a subgroup of $\mathcal{F}$.

Case L' has fusion pattern 302: Then C(L') has type D_4 and is conjugate to a subgroup of H.

Case L' has fusion pattern 361: Then C(L') has type B_3 and is a subgroup of the D_6 factor of the centralizer of a 5G element which, in turn, is conjugate to a subgroup of H.

Case L' has fusion pattern 557: Then C(L') has type A_4 and is a subgroup of the D_5 factor of the connected centralizer of a Dih_{10} subgroup of L' which, in turn, is a subgroup of the D_6 factor of the connected centralizer of a 5G element which, in turn, is conjugate to a subgroup of H.

Case L' has fusion pattern 694: Then C(L') has type F_4, and is conjugate to $\mathcal{F}$. Hence C(L) contains a subgroup of type G_2.

Case L' has fusion pattern 753: Then C(L') has type D_5 which is conjugate to a subgroup of H.

Case L' has fusion pattern 769: Then C(L') has type E_6. So L is conjugate to a subgroup of $\mathcal{E}$ and C contains $C(\mathcal{E})$ which has type A_2.

By exhaustion of cases, the Lemma is proved. ∎

Case 785 (798): 2B, 3A, 5A

By Table 4.5, C has dimension 8. Since by [CoGr '93] (3.2), $C_G(Y)$ has type $A_2 A_2$, C has rank at most 4. There are two reductive group types of rank ≤ 4 and dimension 8, and hence two possibilities for the type of C:

1) C has type $A_1 A_1 T_2$. Then C contains a three dimensional torus and some conjugate of L is therefore a subgroup of H by Lemma 5.23.

2) C has type A_2. By Lemma 7.4, either some conjugate of L is a subgroup of H or some conjugate of L is a subgroup of $\mathcal{E}$. But by Table 6.4, $\mathcal{E}$ does not contain an Alt_5 subgroup with this fusion pattern, so some conjugate of L is a subgroup of H.

In either case, we have that some conjugate of L is a subgroup of H. But by Tables 5.13 and 5.19, there is only one H-conjugacy class of Alt_5 subgroups with this fusion pattern in H and therefore only one G-class of Alt_5 subgroups with this fusion pattern in G. By Lemma 5.35, since the maximal torus of the connected centralizer in H of the Alt_5 subgroups of H with this fusion pattern contains an element of type 2B and has dimension 2, $C_G(L)$ cannot have type $A_1A_1T_2$, so has type A_2.

<u>Case 815 (828)</u>: 2B, 3A, 5B

By Table 4.5, C has dimension 2, so C is a two dimensional torus T, and therefore by Lemma 5.25, L is conjugate to a subgroup of H. By Tables 5.13 and 5.19, there are two classes of embeddings which correspond to this fusion pattern, namely, $5 + 3^2 + 3' + 1^2$, and $5 + 3 + 4^2$, so there are two H-classes of Alt_5 subgroups in H and therefore either one or two G-classes. Suppose L corresponds to the $5 + 3 + 4^2$ class of embeddings. Then by Lemmas 5.32 and 5.34, there is a one-dimensional torus in H which centralizes L. Now the connected centralizer of L is a two dimensional torus, but since only one dimension is visible in H, the other dimension is not in H by Lemmas 5.32 and 5.33 and therefore does not centralize z. The element of order two in the one-dimensional torus in H which centralizes L has type 2B, and so by Lemma 5.35, $\exists g \in G$ such that $L^g \leq H$ and $C_H(L^g)$ contains a 2-dimensional torus. But then by Lemmas 5.32 and 5.34, L^g corresponds to the $5 + 3^2 + 3' + 1^2$ class of embeddings. Hence there is only one G-class of Alt_5 subgroups with this fusion pattern in G.

Case 844: 2B, 3A, 5C

Subgroups of G with this fusion pattern are the ZDC's. We are interested in ZDC's which have elements of orders two and three in their centralizers, because we can gain insight into Alt_5 subgroups of G with fusion pattern 272, and SL(2, 5) subgroups of G with fusion pattern 633.

Suppose L centralizes an element y of order 3. Then by Table 5.38, there is no element of type 3A in C(L) so y has type 3B and L is diagonally embedded in a group of type A_2E_6. Let $C(y) = X_1X_2$ where $X_1 \cong SL(3, \mathbb{C})$ and X_2 has type E_6. Let L_1 and L_2 be the commutator subgroups of the quasiprojections (see Definition 1.28) of L into X_1 and X_2 respectively. So $L_i \cong Alt_5$ for i = 1, 2. Then L_1 has fusion pattern in the class {769, 782} and so is unique up to conjugacy in both G and X_1 (see discussion of case 769 on p. 120) and L_2 has fusion pattern in the class {272, 285} by Table 6.4 since L_2 has a 0-dimensional centralizer in X_2. Let $\phi_1 : L_1 \to L_2$ and $\phi_2 : L_1 \to L_2$ be isomorphisms. Then $\phi_2 \phi_1^{-1}$ is an automorphism of L_2. So any isomorphism $\phi' : L_1 \to L_2$ is of the form $\sigma \circ \phi_1$ where σ is an automorphism of L_2, and therefore L is of the form $\{\ell\sigma(\phi_1(\ell)) \mid \ell \in L_1\}$ for some $\sigma \in Aut(L_2)$. Let $Inn(L_2)$ be the group of inner automorphisms of L_2, $\mathcal{L}^{in}$ the set of all groups of the form $\{\ell\sigma(\phi_1(\ell)) \mid \ell \in L_1, \sigma$ a fixed element of $Inn(L_2)\}$, and $\mathcal{L}^{out}$ the set of all groups of the form $\{\ell\sigma(\phi_1(\ell)) \mid \ell \in L_1, \sigma$ a fixed element of $Aut(L_2)\backslash Inn(L_2)\}$. Let $A_1 := \{\ell\sigma_1(\phi_1(\ell)) \mid \ell \in L_1, \sigma_1$ a fixed element of $Inn(L_2)\}$ and $A_2 := \{\ell\sigma_2(\phi_1(\ell)) \mid \ell \in L_1, \sigma_2$ a fixed element of $Inn(L_2)\}$ be two elements of $\mathcal{L}^{in}$. Then since σ_1 and σ_2 are inner, there exists an element $g \in L_2$ such that $A_1^g = A_2$, so all the elements of $\mathcal{L}^{in}$ are conjugate in G. Now let $B_1 := \{\ell\sigma_1(\phi_1(\ell)) \mid \ell \in L_1, \sigma_1$ a fixed element of $Aut(L_2)\backslash Inn(L_2)\}$ and $B_2 := \{\ell\sigma(\phi_1(\ell)) \mid \ell \in L_1, \sigma_2$ a fixed element of $Aut(L_2)\backslash Inn(L_2)\}$ be two elements of $\mathcal{L}^{out}$. Since $\sigma_1 = \sigma_2\eta$ where $\eta \in Inn(L_2)$, there exists a $g \in L_2$ such that $B_1^g = B_2$, so all the elements of $\mathcal{L}^{out}$ are conjugate in G. It is not clear however whether the elements of $\mathcal{L}^{in}$ are conjugate in G to the elements of $\mathcal{L}^{out}$, so for each class of Alt_5 subgroup of X_2 with fusion pattern in the class {272, 285}, there are either one or two classes of ZDC's with an element of order 3 in the centralizer. Let n be the number of X_2-classes of Alt_5-subgroups of X_2 with fusion pattern 272. Then there are between n and 2n classes of ZDC's with an element of order 3 in the centralizer.

Suppose L centralizes an element x of type 2A. Then L is diagonally embedded in a group of type A_1E_7. Let $C(x) = X_1X_2$ where $X_1 \cong SL(2, \mathbb{C})$ and X_2 has type E_7. Let M_1 and M_2 be quasiprojections of L into X_1 and X_2 respectively. Then M_1 and $M_2 \cong SL(2, 5)$ since X_1 doesn't have any Alt_5-subgroups. So M_1 has fusion pattern in the class {2294} by Table 5.36 and is unique up to conjugacy in X_1, and M_2 is conjugate to a fixed point free $SL(2, 5)$-subgroup of S with central involution z_1. So $C_G(M_2) \cong SL(2, \mathbb{C})$. The only such $SL(2, 5)$ subgroups of G have fusion pattern 633 since $SL(2, 5)$-subgroups of G with fusion patterns 174 and 188 are not conjugate to subgroups of S (see Remark 8.0 p. 140). Let $\phi_1 : M_1 \to M_2$ be an isomorphism. Then any other isomorphism $\phi' : M_1 \to M_2$ is of the form $\sigma \circ \phi_1$ where σ is an automorphism of M_2. By the same arguments as above, for each class of $SL(2, 5)$ subgroups of X_2 with fusion pattern 633, there are either one or two classes of ZDC's with an element of type 2A in the centralizer. Let m be the number of X_2-classes of $SL(2, 5)$ subgroups of X_2 with fusion pattern 633. Then there are between m and 2m classes of ZDC's with an element of type 2A in the centralizer.

We should mention the Borovik group here. By [CoGr '93], the Borovik group contains a group S of type $Alt_5 \times \Sigma_6$ with index 2. Also by [CoGr '93], the Borovik group is unique up to conjugacy (so S is unique up to conjugacy) and the Σ_6 factor contains elements of types 2B, 3B, 4C, 5C, 2A and

6F. Since all ZDC's with elements of type 2B in the centralizer are conjugate to the ZDC in H they are all therefore conjugate to the Alt_5 factor of S. I would like to prove that there is only one class of ZDC's with nontrivial centralizer, i.e. that all such ZDC's are conjugate to the Alt_5 factor of S. If that is true, then m = n = 1 where m and n are as in the above paragraphs.

<u>Case 860 (873)</u>: 2B, 3A, 5D

By Table 4.5, the dimension of C is 3, and since there are two reductive group types of dimension three, there are two possibilities for the type of C:

1) C is a three dimensional torus, and therefore by Lemma 5.23, some conjugate of L is a subgroup of H.

2) C has type A_1. By Lemma 6.5, L is conjugate to a subgroup of H or $\mathcal{A}$, or L has an element of type 3B in its connected centralizer. If L has an element of type 3B in its connected centralizer, it is unique up to conjugacy in G, and since the members of the class of Alt_5-subgroups of H with this fusion pattern have elements of type 3B in their centralizers, L is conjugate to a member of this class. By Table 4.16, L is not conjugate to a subgoup of $\mathcal{A}$, and therefore L is conjugate to a subgroup of H.

Since, by Tables 5.13 and 5.19, there is only one H-class of Alt_5 subgroups with this fusion pattern in H, there is a unique G-class of Alt_5 subgroups of G with this fusion pattern. By Lemma 5.35, since the maximal torus of the connected centralizer in H of the Alt_5 subgroups of H with this fusion pattern contains an element of type 2B and has dimension 1, $C_G(L)$ cannot have type T_3, so has type A_1.

Case 1040: 2B, 3B, 5C

By Table 4.5, the dimension of C is 3. There are two reductive group types of dimension three so there are two possibilities for the type of C:

1) C is a three dimensional torus, so by Lemma 5.23, some conjugate of L is a subgroup of H.

2) C has type A_1. By Lemma 6.5, L is a subgroup of a conjugate of H, $\mathcal{A}$ or L has an element of type 3B in its connected centralizer. But if L has an element of type 3B in its connected centralizer, L is unique up to conjugacy among Alt_5 -subgroups of G with centralizer of rank 1, and therefore L is conjugate to a subgroup of H with fusion pattern 860, a contradiction. Hence L is conjugate to a subgroup of either $\mathcal{A}$ or H. But the connected centralizer of the unique H-class of Alt_5 subgroups of H with this fusion pattern contains an element of type 3A, and therefore, since, by Table 4.16, there is a unique $\mathcal{A}$-class of Alt_5 subgroups of $\mathcal{A}$, the unique (by Tables 5.13 and 5.19) H-class of Alt_5 subgroups of H with this fusion pattern is G-conjugate to the class of Alt_5 subgroups of $\mathcal{A}$ with this fusion pattern.

In any case, L is a subgroup of some conjugate of H. But, by Tables 5.13 and 5.19, there is only one H-class of Alt_5 subgroups of H with this fusion pattern. Hence there is only one G-class of Alt_5 subgroups of G with this fusion pattern. By Lemma 5.35, since the maximal torus of the connected centralizer in H of the Alt_5 subgroups of H with this fusion pattern contains an element of type 2B and has dimension 1, $C_G(L)$ cannot have type T_3, so has type A_1.

Case 1056 (1069): 2B, 3B, 5D

By Table 4.5, C has dimension 6, and therefore by Lemma 5.25, L is conjugate to a subgroup of H. Since, by [CoGr '93] (3.2), $C_G(Y)$ has type A_2G_2, C has rank at most 4. There are two reductive group types of rank ≤ 4 and dimension 6, and hence two possibilities for the type of C: A_1T_3 and A_1A_1. By Tables 5.13 and 5.19, there is only one H-class of Alt_5-subgroups of H with this fusion pattern. Hence there is only one G-class of Alt_5-subgroups of G with this fusion pattern. By Lemma 5.35, since the maximal torus of the connected centralizer in H of the Alt_5 subgroups of H with this fusion pattern contains an element of type 2B and has dimension 2, $C_G(L)$ cannot have type A_1T_3, so has type A_1A_1.

Case 1177 (1190): 2B, 3C, 5A

By Table 4.5, C has dimension 14. Since, by [CoGr '93] (3.2), $C_G(Y)$ has type G_2G_2, C has rank at most 4. Also note that by Table 3.11, $C_G(Z)$ has type D_4. There are three reductive group types of rank ≤ 4 and dimension 14, and hence there are three possibilities for the type of C: $B_2A_1T_1$, $A_2A_1A_1$ and G_2. If C has type $B_2A_1T_1$ or $A_2A_1A_1$, then by Lemma 5.23, L is conjugate to a subgroup of H. But then by Lemmas 5.35, 5.32 and 5.34 and Tables 5.13 and 5.19, C(L) has rank 2, a contradiction, so C has type G_2. Therefore by Lemma 7.4, either some conjugate of L is a subgroup of H, or some conjugate of L is a subgroup of $\mathcal{F}$, and therefore a subgroup of $\mathcal{E}$. Since, by Table 6.4, there are no Alt_5 subgroups of $\mathcal{E}$ with this fusion pattern, some conjugate of L is a subgroup of H. Since, by Tables 5.13 and 5.19, there is exactly one H-class of Alt_5 subgroups of H with this fusion pattern, there is exactly one G-class of Alt_5 subgroups of G with this fusion pattern.

Case 1207 (1220): 2B, 3C, 5B

By Table 4.5, C has dimension 8. Since, by [CoGr '93] (3.2), $C_G(Y)$ has type G_2G_2, C has rank at most 4. There are two reductive group types of rank 4 and dimension 8, and hence there are two possibilities for the type of C:

1) C has type $A_1A_1T_2$. Then C contains a three dimensional torus and some conjugate of L is therefore a subgroup of H by Lemma 5.23.

2) C has type A_2. By Lemma 7.4, some conjugate of L is a subgroup of either H or $\mathcal{E}$. Since, by Table 6.4, there is no Alt_5 subgroup of $\mathcal{E}$ with this fusion pattern, L has a conjugate in H.

In either case L has a conjugate in H. By Tables 5.13 and 5.19, there are two classes of embeddings which correspond to this fusion pattern, namely, $4 + 3^2 + 3' + 1^3$, and $4^3 + 3 + 1$. Hence there are two H-classes of Alt_5 subgroups in H and therefore either one or two G-classes of Alt_5 subgroups with this fusion pattern in G. Suppose L corresponds to the $4^3 + 3 + 1$ class of embeddings. Then by Lemmas 5.32 and 5.34, there is a one-dimensional torus in H which centralizes L. Now the connected centralizer of L contains a two dimensional torus, but since only one dimension is visible in H, the other dimension(s) is not in H and therefore, by Lemmas 5.32 and 5.33 does not centralize z. The element of order two in the one-dimensional torus in H which centralizes L has type 2B, and so, by Lemmas 5.35, 5.32 and 5.34 and Tables 5.13 and 5.19, $\exists g \in G$ such that $L^g \leq H$ and $C_H(L^g)$ has rank 2. But then L^g corresponds to the $4 + 3^2 + 3' + 1^3$ class of embeddings. Hence there is only one G-class of Alt_5 subgroups with this fusion pattern in G. By Lemma 5.35, since the centralizers in H of the H-classes of Alt_5 subgroups of H with this fusion pattern contain an element of type 2B and have dimension at most 2, $C_G(L)$ cannot have type $A_1A_1T_2$, so has type A_2.

Case 1236: 2B, 3C, 5C

By Table 4.5, C has dimension 6, and therefore by Lemma 5.25, L is conjugate to a subgroup of H. Since, by [CoGr '93] (3.2), $C_G(Y)$ has type G_2G_2, C has rank at most 4. There are two reductive group types of rank ≤ 4 and dimension 6, and hence two possibilities for the type of C: A_1T_3 and A_1A_1. Since, by Tables 5.13 and 5.19, there is only one H-class of Alt_5 subgroups in H, there is only one G-class of Alt_5 subgroups of G. By Lemma 5.35, since the maximal torus of the connected centralizer of the Alt_5 subgroups of H with this fusion pattern contains an element of type 2B and has dimension 2, $C_G(L)$ cannot have type A_1T_3, so has type A_1A_1.

Case 1252 (1265): 2B, 3C, 5D

By Table 4.5, C has dimension 9, and therefore by Lemma 5.25, L is conjugate to a subgroup of H. Since, by [CoGr '93] (3.2), $C_G(Y)$ has type G_2G_2, C has rank at most 4. Also note that $C_G(Z)$ has type $B_2B_2B_1$ by Table 3.11. There are two reductive group types of rank ≤ 4 and dimension 9, and hence there are two possibilities for the type of C: A_2T_1 and $A_1A_1A_1$. If C has type A_2T_1, then C embeds in $C_G(Z)$, and the A_2 factor of C embeds into one of the B_2 factors of $C_G(Z)$. The root lattice for A_2 is not a sublattice of the root lattice for B_2 since the determinant of the root lattice of A_2 is 3 while that for B_2 is 2 which contradicts [Gr unp.](8.12). So C has type $A_1A_1A_1$. By Tables 5.13 and 5.19, there are two classes of embeddings which correspond to this fusion pattern, namely, $3^3 + 3' + 1^4$, and $4^2 + 3^2 + 1^2$. Hence, there are two H-classes of Alt_5 subgroups with this fusion pattern in H, and therefore either one or two G-classes of Alt_5 subgroups with this fusion pattern in G. In either case, however, the connected centralizer of an Alt_5 subgroup L which corresponds to one of these classes of embeddings contains an element of type 3A which forces L to be conjugate to a subgroup of $\mathcal{A}$. But by Table 4.16, there is only one $\mathcal{A}$-class of Alt_5 subgroups of $\mathcal{A}$ with this fusion pattern, and therefore there is only one G-class of Alt_5 subgroups with this fusion pattern in G.

Case 1312 (1325): 2B, 3C, 5F

By Table 4.5, C has dimension 28, and therefore by Lemma 5.25, L is conjugate to a subgroup of H. Since, by [CoGr '93] (3.2), $C_G(Y)$ has type G_2G_2, C has rank at most 4. There are two reductive group types with rank ≤ 4 and dimension 28, and hence two possibilities for the type of C: D_4 and G_2G_2. If C has type D_4, then C is a subgroup of one of the factors of $C_G(Y)$. But this is impossible since D_4 acts on an 8 dimensional module, whereas G_2 acts on a 7 dimensional module. Hence, C has type G_2G_2. By Tables 5.13 and 5.19, H has only one H-class of Alt_5 subgroups with this fusion pattern so there is only one G-class of Alt_5 subgroups with this fusion pattern.

Case 1341: 2B, 3C, 5G

By Table 4.5, C has dimension 16, and therefore by Lemma 5.25, L is conjugate to a subgroup of H. Since, by [CoGr '93] (3.2), $C_G(Y)$ has type G_2G_2, C has rank at most 4. Also note that by Table 3.11, $C_G(Z)$ has type A_3A_3. There are four reductive group types of rank ≤ 4 and dimension 16, and hence four possibilities for the type of C:

1) C has type G_2T_2. Then the G_2 factor of C embeds into one of the factors of $C_G(Z)$. But a group of type G_2 acts on a 7 dimensional module, whereas a group of type A_3 acts on a 4 dimensional module. So C doesn't have this type.

2) C has type A_3T_1. Then the A_3 factor of C embeds into one of the factors or a diagonal group of $C_G(Y)$, but a group of type A_3 cannot embed into a group of type G_2. So C doesn't have this type.

3) C has type $B_2A_1A_1$. Then the B_2 factor of C embeds into one of the factors or a diagonal group of $C_G(Y)$, but a group of type B_2 cannot embed into a group of type G_2. So C doesn't have this type.

So C has type A_2A_2. By Tables 5.13 and 5.19, there are two $GL(16, \mathbb{C})$-classes of embeddings which correspond to this fusion pattern, namely, $3^2 + 3'^2 + 1^4$, and 4^4. (Now $3^2 + 3'^2 + 1^4$ is actually an $SO(16, \mathbb{C})$-class of embeddings by Remark 5.10, but at this point, we are unable to say the same for 4^4). Suppose L corresponds to the 4^4 class of embeddings. Then by Lemmas 5.32 and 5.34, there is a two-dimensional torus in H which centralizes L. Now the connected centralizer contains a four-dimensional torus, but since only two dimensions are visible in H, by Lemmas 5.32 and 5.33, the other dimensions are not in H and therefore do not centralize z. Now there is an element of type 2B in the two-dimensional torus in H which centralizes L, and that element is conjugate to z, say by g. Then L^g is in the centralizer of z (i.e. H), and z centralizes a four-dimensional torus in the connected centralizer of L^g so a four dimensional torus in the connected centralizer of L^g is contained in H. But then L^g corresponds to the $3^2 + 3'^2 + 1^4$ class of embeddings. Hence there is only one G-class of Alt_5 subgroups with this fusion pattern in G.

Theorem 7.5 Suppose L is an Alt_5 subgroup of G. Then L is conjugate to every other Alt_5 subgroup of G with the same fusion pattern unless L has fusion pattern 272 or 844. These fusion patterns and the centralizer types of Alt_5 subgroups of G with these fusion patterns are listed in Table 7.6.

Table 7.6 Conjugacy classes of Alt_5 subgroups of G.

These are the fusion patterns of Alt_5 subgroups which are contained in G (= E_8), some important subgroups in which the Alt_5 subgroups are contained, their connected centralizers, and the number of conjugacy classes of each in G.

Fusion pattern		Subgroups of G which contain Alt_5 subgroups of the given fusion pattern	Connected centralizer Type	Number of conjugacy classes in G
272	2A 3B 5D	$\mathcal{A}, \Delta, H, \mathcal{E}$	G_2	≥ 1*
302	2A 3B 5E	$\mathcal{A}, \Delta, H, \mathcal{E}$	D_4	1
361	2A 3B 5G	$H, \mathcal{E}$	B_3	1
557	2A 3C 5G	$\mathcal{A}, H, \mathcal{E}$	A_4	1
694	2A 3D 5E	$\Omega, H, \mathcal{E}$	F_4	1
753	2A 3D 5G	$\mathcal{A}, \Omega, H, \mathcal{E}$	D_5	1
769	2A 3D 5H	$\mathcal{A}, \Delta, \Omega, H, \mathcal{E}$	E_6	1
785	2B 3A 5A	Δ, H	A_2	1
815	2B 3A 5B	Δ, H	T_2	1
844	2B 3A 5C	Δ, H	1	≥ 1
860	2B 3A 5D	H	A_1	1
1040	2B 3B 5C	$\mathcal{A}, H$	A_1	1
1056	2B 3B 5D	$\mathcal{A}, H$	$A_1 A_1$	1
1177	2B 3C 5A	Ω, H	G_2	1
1207	2B 3C 5B	Ω, H	A_2	1
1236	2B 3C 5C	$\mathcal{A}, \Omega, H$	$A_1 A_1$	1
1252	2B 3C 5D	$\mathcal{A}, \Omega, H$	$A_1 A_1 A_1$	1
1312	2B 3C 5F	$\mathcal{A}, \Delta, \Omega, H, \mathcal{E}$	$G_2 G_2$	1
1341	2B 3C 5G	$\mathcal{A}, \Delta, \Omega, H, \mathcal{E}$	$A_2 A_2$	1

*We actually know more than this. See discussion of case 844 on p. 124.

Remark 7.7 It is clear that for each conjugacy class K of Alt_5 subgroups which have nonrational elements of order 5, that there are two nonconjugate classes of embeddings of Alt_5 subgroups of K. As for those classes which have rational elements of order 5, all of them have representatives in H (with the possible exception of classes of ZDC's). Suppose K is an Alt_5 conjugacy class where the elements of K have rational elements of order 5. Then by Theorem 7.5, K consists of all Alt_5 subgroups of G with the given fusion pattern so if there is any element of K afforded by an embedding of Alt_5 into H which is unchanged by an outer twist and if there is an element of H which induces that outer twist, then there is only one class of embeddings affording elements of K. Such an element exists in H if any of the irreducible constituents are odd. We produce such an embedding for each rational class K (except for the classes of ZDC's).

$361: 5 + 4 + 1^7$

$557: 5^2 + 1^6$

$753: 3 + 3' + 1^{10}$

$1040: 5^2 + 3 + 3'$

$1236: 4^2 + 3 + 3' + 1^2$

$1341: 3^2 + 3'^2 + 1^4$

Chapter 8
Conjugacy classes of SL(2, 5) subgroups of G

We now begin to analyze each of the SL(2, 5) cases. In what follows, let M be an SL(2, 5) subgroup of G and $C = C_G(M)$.

<u>Case 3</u>: 2A, 3A, 4A, 5A, 6C, 10N

By Table 4.9, C has dimension 11 and therefore M is conjugate to a subgroup of H by Lemma 5.24, and C has one of the following types: B_2T_1, $A_1A_1A_1T_2$, A_2A_1, A_2T_3, $A_1A_1T_5$. By Table 5.36, there are two GL(16, $\mathbb{C}$)-classes of embeddings which correspond to this fusion pattern, namely, $2^2 + 3 + 3'2 + 1^3$, and $2^2 + 4^3$ (Now $2^2 + 3 + 3'2 + 1^3$ actually corresponds to an SO(16, $\mathbb{C}$)-class of embeddings by Remark 5.10, but at this point we are unable to say the same for $2^2 + 4^3$). Suppose M corresponds to the $2^2 + 4^3$ class of embeddings. Then by Lemmas 5.32 and 5.34, there is a two-dimensional torus T in H which centralizes M. Now the connected centralizer in G of M contains a three-dimensional torus and T contains an element of type 2B, so by Lemma 5.35, $\exists g \in G$ such that $M^g \subseteq H$ and $C_H(M^g)$ contains a three-dimensional torus. But then M^g corresponds to the $2^2 + 3 + 3'2 + 1^3$ class of embeddings, so there is only one G-class of SL(2, 5) subgroups with this fusion pattern. Moreover, by Lemma 5.35, C can have rank at most 3 so has type B_2T_1 or A_2A_1.

<u>Case 19</u>: 2A, 3A, 4A, 5B, 6C, 10D

By Table 4.9, C has dimension 11 and therefore M is conjugate to a subgroup of H by Lemma 5.24, and C has one of the following types: B_2T_1, $A_1A_1A_1T_2$, A_2A_1, A_2T_3, $A_1A_1T_5$. By Table 5.36, Remark 5.10 and Lemma 5.4, there is only one H-class of SL(2, 5) subgroups with this fusion pattern in H, namely, the class which corresponds to $2^2 + 3^3 + 1^3$, and therefore only one G-class of such subgroups in G. Moreover, by Lemma 5.35, C can have rank at most 3 so has type A_2A_1.

Case 22: 2A, 3A, 4A, 5B, 6C, 10Z

By Table 4.9, C has dimension 7 and therefore M is conjugate to a subgroups of H by Lemma 5.24, and C has one of the following types: $A_1A_1T_1$, A_1T_4, T_7. By Table 5.36, Remark 5.10 and Lemma 5.4, there are three H-classes of SL(2, 5) subgroups of H with this fusion pattern, namely, those classes which correspond to $2^2 + 3^2 + 3' + 1^3$, $2^2 + 4 + 3'^2 + 1^2$ and $2^2 + 4^2 + 3 + 1$. The centralizers of the members of the classes corresponding to $2^2 + 3^2 + 3' + 1^3$ and $2^2 + 4 + 3'^2 + 1^2$ contain elements of type 3A, and therefore the members of those two classes are conjugate to subgroups of $\mathcal{A}$. Since there is only one $\mathcal{A}$-class of SL(2, 5) subgroups with this fusion pattern in $\mathcal{A}$, these two H-classes fuse into one G-class. Now suppose M corresponds to the $2^2 + 4^2 + 3 + 1$ class of embeddings. Then by Lemmas 5.32 and 5.34 there is a two dimensional torus T in H which centralizes M. Now the connected centralizer in G of M contains a three dimensional torus and T contains an element of type 2B, so by Lemma 5.35, $\exists g \in G$ such that $M^g \subseteq H$ and $C_H(M^g)$ contains a three dimensional torus. But then M^g corresponds to one of the other two H-classes of embeddings (which form a single G-class). So there is only one G-class of SL(2, 5) subgroups with this fusion pattern. Moreover, by Lemma 5.35, C can have rank at most 3 so has type $A_1A_1T_1$.

Case 37: 2A, 3A, 4A, 5C, 6C, 10U

By Table 4.9, C has dimension 9 and therefore M is conjugate to a subgroup of H by Lemma 5.24, and C has one of the following types: $A_1A_1A_1$,

A_2T_1, $A_1A_1T_3$, A_1T_6. By Table 5.36, Remark 5.10 and Lemma 5.4, there are two H-classes of SL(2, 5) subgroups of H with this fusion pattern, namely the classes corresponding to $2^2 + 3'^3 + 1^3$ and $2^2 + 4 + 3^2 + 1^2$. Now the centralizers of the members of both H-classes contain elements of type 3A, and therefore the members of both H-classes are conjugate to subgroups of $\mathcal{A}$. Since there is only one $\mathcal{A}$-class of SL(2, 5) subgroups with this fusion pattern in $\mathcal{A}$, there is only one G-class of SL(2, 5) subgroups with this fusion pattern in G. Moreover, by Lemma 5.35, C can have rank at most 3 so has type $A_1A_1A_1$ or A_2T_1.

Case 57: 2A, 3A, 4A, 5D, 6C, 10HH

By Table 4.9, C has dimension 6 and therefore M is conjugate to a subgroup of H by Lemma 5.24, and C has one of the following types: A_1A_1, A_1T_3, T_6. By Lemma 5.35, C has rank at most 2 so has type A_1A_1. If C(M) contains a 2A-pure fours-group, then M is conjugate to a subgroup of $\mathcal{E}$ and then C(M) $\supseteq$ a subgroup of type A_2 which is impossible. Hence C(M) contains an element of type 2B and M is conjugate to a subgroup of H. By Table 5.36, Remark 5.10 and Lemma 5.4, there are two H-classes of SL(2, 5) subgroups with this fusion pattern in H, namely, the classes corresponding to $2^2 + 4 + 3 + 3' + 1^2$ and $2^2 + 4^2 + 3' + 1$. There is therefore either one or two G-classes of SL(2, 5) subgroups in G with this fusion pattern. Suppose M corresponds to the $2^2 + 4^2 + 3' + 1$ class of embeddings. Then the toral nines-group in C(M)° contains an element of type 3C, so M is conjugate to a subgroup of the standard subgroup of H of type $4D_7$. But by Table 5.13, there is only one class of embeddings of SL(2, 5)-subgroups of G with this fusion pattern in that subgroup, namely the class of embeddings which corresponds to $2^2 + 4 + 3 +$

$3' + 1^2$. Hence the two H-classes of SL(2, 5)-subgroups fuse into one G-class so there is only one G-class of subgroups with this fusion pattern in G.

Case 152: 2A, 3A, 4D, 5A, 6G, 10L

By Table 4.9, C has dimension 9 and therefore M is conjugate to a subgroup of H by Lemma 5.24, and C has one of the following types: $A_1A_1A_1$, A_2T_1, $A_1A_1T_3$, A_1T_6. By Table 5.36, there are two GL(16, $\mathbb{C}$)-classes of embeddings which correspond to this fusion pattern, namely, $2^2 + 5 + 3' + 1^4$, and $4_f^2 + 2^2 + 4$ (Now $2^2 + 5 + 3' + 1^4$ actually corresponds to an SL(16, $\mathbb{C}$)-class of embeddings by Remark 5.10, but at this point we are unable to say the same for $4_f^2 + 2^2 + 4$). Suppose M corresponds to the $4_f^2 + 2^2 + 4$ class of embeddings. Then by Lemmas 5.32 and 5.34, there is a two dimensional torus T in H which centralizes M. Now the connected centralizer in G of M contains a three dimensional torus and T contains an element of type 2B, so by Lemma 5.35, $\exists g \in G$ such that $M^g \subseteq H$ and $C_H(M^g)$ contains a three dimensional torus. But then M^g corresponds to the $2^2 + 5 + 3' + 1^4$ class of embeddings, so there is only one G-class of SL(2, 5) subgroups with this fusion pattern. Moreover, by Lemma 5.35, C can have rank at most 3 so C has type $A_1A_1A_1$ or A_2T_1. By Table 5.36 entry 27, C contains a subgroup of type A_1A_1 and therefore has type $A_1A_1A_1$.

Case 170: 2A, 3A, 4D, 5B, 6G, 10P

By Table 4.9, C has dimension 9 and therefore M is conjugate to a subgroup of H by Lemma 5.24, and C has one of the following types: $A_1A_1A_1$, A_2T_1, $A_1A_1T_3$, A_1T_6. By Table 5.36, Remark 5.10 and Lemma 5.4, there are two H-classes of SL(2, 5) subgroups with this fusion pattern in H, namely, $2^2 + 5 + 3 + 1^4$, and $4_f^2 + 2^2 + 3 + 1$. Suppose M corresponds to the $4_f^2 + 2^2 + 3 + 1$

class of embeddings. Then by Lemmas 5.32 and 5.34, there is a two dimensional torus T in H which centralizes M. Now the connected centralizer in G of M contains a three dimensional torus and T contains an element of type 2B, so by Lemma 5.35, $\exists g \in G$ such that $M^g \subseteq H$ and $C_H(M^g)$ contains a three dimensional torus. But then M^g corresponds to the $2^2 + 5 + 3 + 1^4$ class of embeddings, so there is only one G-class of SL(2, 5) subgroups with this fusion pattern. Moreover, by Lemma 5.35, C can have rank at most 3 so has type $A_1A_1A_1$ or A_2T_1. By Table 5.36 entry 26, C contains a subgroup of type A_1A_1 and therefore has type $A_1A_1A_1$.

Case 174: 2A, 3A, 4D, 5B, 6G, 10JJ

By Table 4.9, C has dimension 3 and has one of the following types: A_1, T_3. If C has type T_3, then M is conjugate to a subgroup of H. But then by Lemmas 5.35, 5.32 and 5.34 and Table 5.36, C has rank 1. So C has type A_1. Now let y be the element of order three in C°. Then $M \leq C(y)$. If y has type 3A, then M is conjugate to a subgroup of $\mathcal{A}$, but there are no SL(2, 5)-subgroups of $\mathcal{A}$ with this fusion pattern. If y has type 3D, then M is conjugate to a subgroup of $\mathcal{S}$. By Table 1.19 the elements of order 6 have type 6J[$\mathcal{S}$] or 6K[$\mathcal{S}$] and therefore the elements of order 2 have type 2B[$\mathcal{S}$] and the elements of order 3 have type 3A[$\mathcal{S}$]. By Table 1.20, if g is an element of M of order 10, then $\xi(g) = 2 + 6\tau$ and g^2 has type 5B[$\mathcal{S}$], $\xi(g) = 2 - 4\tau$ and g^2 has type 5C[$\mathcal{S}$] or $\xi(g) = 22 - 14\tau$ and g^2 has type 5B[$\mathcal{S}$]. So $(\xi, 1) \in \{10, 14, 12, 4, 8, 6, 26, 30, 28, 20, 24, 22\}$, and $(\xi, 1) > (\chi, 1) = 3$ which is impossible. Therefore M is conjugate to a subgroup of H or there is an element of type 3B in C. Suppose there is an element of type 3B in C. Then, as C has rank 1, M is diagonally embedded into C(y), a group of type A_2E_6. Since y is contained in the 1-dimensional torus T of C(M)° and therefore centralizes T, $T \leq C(y)$. Let π_1 be a quasiprojection of M into the A_2 factor X of C(y), π_2 the corresponding quasi-projection of M into the E_6 factor Y of C(y). Then $\pi_1(M)$ and $\pi_2(M)$ are nontrivial quotients of M and therefore are isomorphic to either Alt_5 or SL(2, 5), but it is not the case that $\pi_1(M) \cong \pi_2(M) \cong Alt_5$.

<u>Case 188</u>: 2A, 3A, 4D, 5C, 6G, 10KK

By Table 4.9, C has dimension 3 and has one of the following types: A_1, T_3. If C has type T_3, then M is conjugate to a subgroup of H. But then by Lemmas 5.35, 5.32 and 5.34 and Table 5.36, C has rank 1. So C has type A_1. Now let y be the element of order three in C°. Then $M \le C(y)$. If y has type 3A, then M is conjugate to a subgroup of $\mathcal{A}$, but there are no SL(2, 5)-subgroups of $\mathcal{A}$ with this fusion pattern. If y has type 3D, then M is conjugate to a subgroup of $\mathcal{S}$. By Table 1.19 the elements of order 6 have type 6J[$\mathcal{S}$] or 6K[$\mathcal{S}$] and therefore the elements of order 2 have type 2B[$\mathcal{S}$] and the elements of order 3 have type 3A[$\mathcal{S}$]. By Table 1.20, if g is an element of M of order 10, then $\xi(g) = 1 - 2\tau$, $11 - 2\tau$ or $29 - 18\tau$ and g^2 has type 5D[$\mathcal{S}$]. So $(\xi, 1) \varepsilon \{10, 12, 14, 4, 6, 8, 26, 28, 30, 20, 22, 24\}$, and $(\xi, 1) > (\chi, 1) = 3$ which is impossible. Therefore M is conjugate to a subgroup of H or there is an element of type 3B in C.

Remark 8.0 By the arguments above, SL(2, 5) subgroups of G with fusion patterns 174 and 188 are not conjugate to subgroups of $\mathcal{S}$. The element of order 3 in the connected centralizers of these groups has type 3B so such an SL(2, 5) subgroup of G is conjugate to a subgroup of G of type A_2E_6.

<u>Case 210</u>: 2A, 3A, 4D, 5D, 6G, 10AAA

By Table 4.9, C has dimension 6 and therefore M is conjugate to a subgroup of H by Lemma 5.24 and C has one of the following types: A_1A_1,

$A_1T_3T_6$. By Lemma 5.35, C has rank at most 2 so has type A_1A_1. If C(M) contains a 2A-pure fours-group, then M is conjugate to a subgroup of $\mathcal{E}$ and then C(M) $\supseteq$ a subgroup of type A_2 which is impossible. Hence C(M) contains an element of type 2B and M is conjugate to a subgroup of H. By Table 5.36, Remark 5.10 and Lemma 5.4, there are two H-classes of SL(2, 5) subgroups with this fusion pattern in H, namely those corresponding to $2^2 + 5 + 4 + 1^3$, and $4_f^2 + 2^2 + 3' + 1$ and therefore either one or two G-classes of SL(2, 5) subgroups with this fusion pattern in G. Suppose M corresponds to the $4_f^2 + 2^2 + 3' + 1$ class of embeddings. Then the toral nines-group in $C(M)^\circ$ contains an element of type 3C, so M is conjugate to a subgroup of the standard subgroup of H of type $4D_7$. But by Table 5.13, there is only one class of embeddings of SL(2, 5)-subgroups of G with this fusion pattern in that subgroup, namely, the class of embeddings which corresponds to $2^2 + 5 + 4 + 1^3$. Hence the two H-classes of SL(2, 5)-subgroups fuse into one G-class so there is only one G-class of subgroups with this fusion pattern in G.

Case 598: 2A, 3B, 4A, 5B, 6F, 10Z

By Table 4.9, C has dimension 4 and therefore M is conjugate to a subgroup of H by Lemma 5.24, and C has one of the following types: A_1T_1, T_4. By Table 5.36, Remark 4.21 and Lemma 5.4, there are two H-classes of SL(2, 5) subgroups with this fusion pattern in H, namely the class which corresponds to $2^2 + 5 + 3'^2 + 1$, and the class which corresponds to $2^2 + 5 + 4 + 3$. Suppose M corresponds to the $2^2 + 5 + 4 + 3$ class of embeddings. Then by Lemmas 5.32 and 5.34 there is a one-dimensional torus in H which centralizes M. Now the connected centralizer in G of M contains a two-dimensional torus, and since a 2A-pure fours-group in C(M) would force M to be conjugate to a subgroup of $\mathcal{E}$ (and then C(M) would have dimension ≥ 8), there is an element of type 2B in C. Hence, by Lemma 5.35, $\exists g \in G$ such that $M^g \subseteq H$ and $C_H(M^g)$ contains a two dimensional torus. But then M^g corresponds to the $2^2 + 5 + 3'2 + 1$ class of embeddings, so there is only one G-class of SL(2, 5) subgroups with this fusion pattern. Moreover, by Lemma 5.35, C can have rank at most 2 so has type A_1T_1.

<u>Case 613</u>: 2A, 3B, 4A, 5C, 6F, 10U

By Table 4.9, C has dimension 6 and therefore M is conjugate to a subgroup of H by Lemma 5.24 and C has one of the following types: A_1A_1, A_1T_3, T_6. By Table 5.36, Remark 5.10 and Lemma 5.4, there is only one H-class of SL(2, 5) subgroups with this fusion pattern in H, namely the class which corresponds to $2^2 + 5 + 3^2 + 1$, and therefore there is only one G-class. Moreover, by Lemma 5.35, C can have rank at most 2 so has type A_1A_1.

<u>Case 633</u>: 2A, 3B, 4A, 5D, 6F, 10HH

By Table 4.9, C has dimension 3 and has one of the following types: A_1, T_3. If C has type T_3, then M is conjugate to a subgroup of H. But then by Lemmas 5.35, 5.32 and 5.34 and Table 5.36, C has rank 1. So C has type A_1.

<u>Case 750</u>: 2A, 3B, 4D, 5B, 6O, 10JJ

By Table 4.9, C has dimension 6 and therefore M is conjugate to a subgroup of H by Lemma 5.24 and C has one of the following types: A_1A_1, A_1T_3, T_6. By Table 5.36, Remark 5.10 and Lemma 5.4, there is only one H-class of SL(2, 5) subgroups in H with this fusion pattern, namely the class which corresponds to $2^2 + 3 + 3'3$, and therefore only one G-class. Moreover, by Lemma 5.35, C can have rank at most 2 so has type A_1A_1.

Case 764: 2A, 3B, 4D, 5C, 6O, 10KK

By Table 4.9, C has dimension 6 and therefore M is conjugate to a subgroup of H by Lemma 5.24, and C has one of the following types: A_1A_1, A_1T_3, T_6. By Table 5.36, Remark 5.10 and Lemma 5.4, there is only one H-class of SL(2, 5) subgroups in H with this fusion pattern, namely, the class which corresponds to $2^2 + 3^3 + 3'$, and therefore only one G-class. Moreover, by Lemma 5.35, C can have rank at most 2 so has type A_1A_1.

Case 785: 2A, 3B, 4D, 5D, 6O, 10ZZ

By Table 4.9, C has dimension 17 and therefore M is conjugate to a subgroup of H by Lemma 5.24 and C has one of the following types: G_2A_1, G_2T_3, $B_2A_1A_1T_1$, $B_2A_1T_4$, $A_2A_2T_1$, $A_2A_1A_1A_1$, $A_2A_1A_1T_3$, $A_1^5 T_2$. By Table 5.36, Remark 5.10 and Lemma 5.4, there is only one H-class of SL(2, 5) subgroups in H with this fusion pattern, namely the class which corresponds to $2^2 + 3^4$, and therefore only one G-class. Moreover, by Lemma 5.35, C can have rank at most 3 so has type G_2A_1.

Case 786: 2A, 3B, 4D, 5D, 6O, 10AAA

By Table 4.9, C has dimension 9 and therefore M is conjugate to a subgroup of H by Lemma 5.24, and C has one of the following types: $A_1A_1A_1$, A_2T_1, $A_1A_1T_3$, A_1T_6. By Table 5.36, Remark 5.10 and Lemma 5.4, there are two H-classes of SL(2, 5) subgroups in H with this fusion pattern, namely the classes which correspond to $2^2 + 5^2 + 1^2$, and $2^2 + 3^2 + 3'^2$. By Lemmas 5.32 and 5.34, there is a three dimensional torus T in H which centralizes M (and T has maximal dimension for a centralizing torus). If M corresponds to the $2^2 + 3^2 + 3'^2$ class of embeddings, then there is a 3B-pure nines-group E in C(M). Since, by Lemma 5.35, C can have rank at most 3, C(E) has type A_2^4 by

[Gr '91](11.4), that is, M is conjugate to a subgroup of Δ. Suppose M_1 is an SL(2, 5) subgroup of H which corresponds to the $2^2 + 5^2 + 1^2$ class of embeddings. Then $C(M_1)$ does not contain any 3B-pure nines-groups, so M_1 is not conjugate to a subgroup of Δ, and therefore M_1 is not conjugate to M. So we have exactly two G-classes of SL(2, 5) subgroups with this fusion pattern in G. Since, by Lemma 5.35, C has rank at most 3, C has type $A_1A_1A_1$ or A_2T_1. It is clear that for the class of subgroups which have a representative in Δ that C has type A_2T_1.

Case 800: 2A, 3B, 4D, 5E, 6O, 10WW

By Table 4.9, C has dimension 17 and therefore M is conjugate to a subgroup of H by Lemma 5.24, and C has one of the following types: G_2A_1, G_2T_3, $B_2A_1A_1T_1$, $B_2A_1T_4$, $A_2A_2T_1$, $A_2A_1A_1A_1$, $A_2A_1A_1T_3$, $A_1^5 T_2$. By Table 5.36, Remark 5.10 and Lemma 5.4, there is only one H-class of SL(2, 5) subgroups in H with this fusion pattern, namely, $2^2 + 3'^4$, and therefore only one G-class. Moreover, by Lemma 5.35, C can have rank at most 3 so has type G_2A_1.

Case 934: 2A, 3B, 4E, 5D, 6P, 10YY

By Table 4.9, C has dimension 39 and therefore M is conjugate to a subgroup of H by Lemma 5.24, and C has one of the following types: B_4A_1, B_4T_3, $A_5A_1T_1$, $D_4B_2T_1$, $D_4A_2A_1$, A_4A_3, $A_4G_2T_1$, $B_3A_3A_1$, $B_3G_2A_1T_1$, $B_3B_2A_2$. By Table 5.36, Remark 5.10 and Lemma 5.4, there are three H-classes of SL(2, 5) subgroups with this fusion pattern in H, namely, $2^4 + 2'^2 + 1^4$, $2^2 + 3' + 1^9$ and $2^2 + 4 + 1^8$. The centralizers of the members of each of the three classes contain elements of type 3A, and therefore the members of each of the three classes are conjugate to subgroups of $\mathcal{A}$. Since there is only one $\mathcal{A}$-class of SL(2, 5) subgroups with this fusion pattern in $\mathcal{A}$, there is only one G-class of

SL(2, 5) subgroups with this fusion pattern in G. Moreover, by Lemma 5.35, C can have rank at most 5 so has type B_4A_1.

Case 951: 2A, 3B, 4E, 5E, 6P, 10BBB

By Table 4.9, C has dimension 55 and therefore M is conjugate to a subgroup of H by Lemma 5.24, and C has one of the following types: B_5, F_4A_1, F_4T_3, D_5B_2. By Table 5.36, Remark 5.10 and Lemma 5.4, there are two H-classes of SL(2, 5) subgroups with this fusion pattern in H, namely $2^6 + 1^4$ and $2^2 + 3 + 1^9$. The centralizers of the members of both classes contain elements of type 3A, and therefore the members of both classes are conjugate to subgroups of $\mathcal{A}$. Since there is only one $\mathcal{A}$-class of SL(2, 5) subgroups with this fusion pattern in $\mathcal{A}$, there is only one G-class of SL(2, 5) subgroups with this fusion pattern in G. Moreover, since, by [Gr '91] (Proposition 9.5(i)), the centralizer of a quaternion subgroup of M has type F_4T_1, C has type F_4A_1.

Case 1310: 2A, 3C, 4D, 5A, 6R, 10L

By Table 4.9, C has dimension 21 and therefore M is conjugate to a subgroup of H by Lemma 5.24, and C has one of the following types: B_3, C_3, $A_3A_1A_1$, $A_3A_1T_3$, $G_2A_1A_1T_1$, $G_2A_1T_4$, $B_2B_2T_1$, $B_2A_2T_3$, $B_2A_1A_1A_1T_2$, $A_2A_2A_1T_2$, $A_2A_1A_1A_1A_1T_1$, A_1^7. By Lemma 5.35, C has rank at most 3 so has type B_3 or C_3. Hence C(M) contains an element of type 2B and M is conjugate to a subgroup of H. By Table 5.36, there are three GL(16, $\mathbb{C}$)-classes of SL(2, 5) subgroups with this fusion pattern in H, namely, the classes which correspond to $2^2 + 4 + 3 + 1^5$, $2^6 + 3' + 1$ and $2^4 + 2'^2 + 4$ (Now $2^2 + 4 + 3 + 1^5$ and $2^6 + 3' + 1$ actually correspond to an SL(16, $\mathbb{C}$)-class of embeddings by Remark 5.10, but at this point we are unable to say the same for $2^4 + 2'^2 + 4$). Suppose M corresponds to the $2^6 + 3' + 1$ class of embeddings. Then C(M) contains an element of type 3C. Similarly, if M_1 is an SL(2, 5) subgroup of H which corresponds to the GL(16, $\mathbb{C}$)-class of embeddings $2^4 + 2'2 + 4$, $C(M_1)$ contains an element of type 3C. Since both $C(M)$ and $C(M_1)$ contain elements of type 3C, so both M and M_1 are conjugate to subgroups of the standard subgroup of H of type $4D_7$. But By Table 5.13, there is only one class of embeddings of SL(2, 5)-subgroups of G with this fusion pattern in that subgroup, namely the class of SL(2, 5)- subgroups corresponding to the $2^2 + 4 + 3 + 1^5$ class of embeddings. Hence the all the the H-classes of SL(2, 5)-subgroups of H fuse into one G-class so there is only one G-class of subgroups with this fusion pattern in G.

Case 1328: 2A, 3C, 4D, 5B, 6R, 10P

By Table 4.9, C has dimension 21 and therefore M is conjugate to a subgroup of H by Lemma 5.24, and C has one of the following types: B_3, $A_3A_1A_1$, $A_3A_1T_3$, $G_2A_1A_1T_1$, $G_2A_1T_4$, $B_2B_2T_1$, $B_2A_2T_3$, $B_2A_1A_1A_1T_2$, $A_2A_2A_1T_2$, $A_2A_1A_1A_1A_1T_1$, A_1^7. By Table 5.36, Remark 5.10 and Lemma 5.4, there are two H-classes of SL(2, 5) subgroups with this fusion pattern in H, namely, $2^2 + 4 + 3 + 1^5$ and $2^4 + 2'2 + 3 + 1$, and therefore either one or two G-classes in G. Moreover, by Lemma 5.35, C can have rank at most 3 and so has type B_3. Suppose M corresponds to the $2^4 + 2'2 + 3 + 1$ class of embeddings. Then $C(M)$ contains an element of type 3C, so M is conjugate to a subgroup of the standard subgroup of H of type $4D_7$. But by Table 5.13, there is only one class of embeddings of SL(2, 5)-subgroups of G with this fusion pattern in that subgroup, namely the class of embeddings which corresponds to $2^2 + 4 + 3 + 1^5$. Hence the two H-classes of SL(2, 5)-subgroups fuse into one G-class so there is only one G-class of subgroups with this fusion pattern in G.

Case 1368: 2A, 3C, 4D, 5D, 6R, 10AAA

By Table 4.9, C has dimension 18 and therefore M is conjugate to a subgroup of H by Lemma 5.24 and C has one of the following types: A_3A_1, A_3T_3, $G_2A_1T_1$, G_2T_4, B_2A_2, $B_2A_1A_1T_2$, $B_2A_1T_5$, $A_2A_2T_2$, $A_2A_1A_1A_1T_1$, $A_2A_1A_1T_3$, A_1^6, $A_1^5 T_3$. By Table 5.36, Remark 5.10 and Lemma 5.4, there are three H-classes of SL(2, 5) subgroups with this fusion pattern in H, namely the classes which correspond to $2^2 + 3 + 3' + 1^6$, $2^2 + 4^2 + 1^4$ and $2^4 + 2'^2 + 3' + 1$. The centralizers of the members of the classes $2^2 + 3 + 3' + 1^6$ and $2^2 + 4^2 + 1^4$ contain elements of type 3A, and therefore the members of these classes are conjugate to subgroups of $\mathcal{A}$. Since there is only one $\mathcal{A}$-class of SL(2, 5) subgroups with this fusion pattern in $\mathcal{A}$, these two H-classes fuse into one G-class. Suppose M corresponds to the $2^4 + 2'^2 + 3' + 1$ class of embeddings. Then by Lemmas 5.32 and 5.34, there is a three-dimensional torus T in H which centralizes M. Now the connected centralizer in G of M contains a four-dimensional torus and T contains an element of type 2B, so by Lemma 5.35, $\exists g \in G$ such that $M^g \subseteq H$ and $C_H(M^g)$ contains a four-dimensional torus. But then M^g corresponds to one of the other two classes of embeddings (which are fused in G), so there is only one G-class of SL(2, 5) subgroups with this fusion pattern in G. Moreover, by Lemma 5.35, C can have rank at most 4 so has type A_3A_1 or B_2A_2. By Table 5.36 entry 21, C contains a subgroup of type A_3 and therefore has type A_3A_1.

Case 1401: 2A, 3C, 4D, 5F, 6R, 10EEE

By Table 4.9, C has dimension 35 and therefore M is conjugate to a subgroup of H by Lemma 5.24, and C has one of the following types: A_5, $D_4A_1A_1T_1$, $A_4B_2T_1$, $A_4A_2A_1$, B_3G_2, $B_3B_2A_1T_1$, $B_3A_2A_1A_1$, $A_3G_2A_1A_1$, $G_2G_2A_1A_1T_1$. By Table 5.36, there are two GL(16, $\mathbb{C}$)-classes of embeddings which correspond to this fusion pattern, namely, $2^2 + 3'2 + 1^6$ and $2^6 + 4$ (Now $2^2 + 3'2 + 1^6$ actually corresponds to an SL(16, $\mathbb{C}$)-class of embeddings by Remark 5.10, but at this point we are unable to say the same for $2^6 + 4$). Suppose M corresponds to the $2^6 + 4$ class of embeddings. Then by Lemmas 5.32 and 5.34, there is a three-dimensional torus T in H which centralizes M. Now the connected centralizer in G of M contains a five-dimensional torus and T contains an element of type 2B, so by Lemma 5.35, $\exists g \in G$ such that $M^g \subseteq H$ and $C_H(M^g)$ contains a five-dimensional torus. But then M^g corresponds to the $2^2 + 3'2 + 1^6$ class of embeddings, so there is only one G-class of SL(2, 5) subgroups with this fusion pattern. Moreover, by Lemma 5.35, C can have rank at most 5 so has type A_5 or B_3G_2.

Case 1419: 2A, 3C, 4D, 5G, 6R, 10DDD

By Table 4.9, C has dimension 35 and therefore M is conjugate to a subgroup of H by Lemma 5.24, and C has one of the following types: A_5, $D_4A_1A_1T_1$, $A_4B_2T_1$, $A_4A_2A_1$, B_3G_2, $B_3B_2A_1T_1$, $B_3A_2A_1A_1$, $A_3G_2A_1A_1$, $G_2G_2A_1A_1T_1$. By Table 5.36, Remark 5.10 and Lemma 5.4, there are two H-classes of SL(2, 5) subgroups with this fusion pattern in H, namely, $2^2 + 3^2 + 1^6$ and $2^6 + 3 + 1$. Suppose M corresponds to the $2^6 + 3 + 1$ class of embeddings. Then by Lemmas 5.32 and 5.34, there is a three-dimensional torus T in H which centralizes M. Now the connected centralizer in G of M contains a five-dimensional torus and T contains an element of type 2B, so by Lemma 5.35, $\exists g \in G$ such that $M^g \subseteq H$ and $C_H(M^g)$ contains a five-dimensional torus. But then M^g corresponds to the $2^2 + 3^2 + 1^6$ class of embedings, so there is only one G-class of SL(2, 5) subgroups with this fusion pattern. Moreover, by Lemma 5.35, C can have rank at most 5 so has type A_5 or B_3G_2.

Case 1504: 2A, 3C, 4E, 5D, 6L, 10YY

By Table 4.9, C has dimension 24 and therefore M is conjugate to a subgroup of H by Lemma 5.24 and C has one of the following types: A_4, B_3A_1, B_3T_3, $A_3A_2T_1$, $A_3A_1A_1A_1$, $A_3A_1A_1T_3$, G_2B_2, $G_2A_2T_2$, $G_2A_1A_1A_1T_1$, $G_2A_1A_1T_4$, $B_2B_2A_1T_1$, $B_2B_2T_4$, $B_2A_2A_1A_1$, $B_2A_2A_1T_3$, $B_2A_1^4T_2$, $A_2A_2A_2$, $A_2A_2A_1A_1T_2$, A_1^8.

By Table 5.36, Remark 5.10 and Lemma 5.4, there are two H-classes of SL(2, 5) subgroups of H with this fusion pattern, namely, those classes which correspond to $4_f 2 + 2^2 + 1^4$ and $2^2 + 5 + 1^7$. The centralizers of the members of both of these classes contains elements of type 3A, and therefore the members of those two classes are conjugate to subgroups of $\mathcal{A}$. Since there is only one $\mathcal{A}$-class of SL(2, 5) subgroups with this fusion pattern in $\mathcal{A}$, these two H-classes fuse into one G-class. Moreover, by Lemma 5.35, C can have rank at most 4 so has type A_4, B_3A_1 or B_2G_2. By Table 5.36 entry 19, C contains a subgroup of type B_3 and therefore has type B_3A_1.

Case 1556: 2A, 3C, 4E, 5G, 6L, 10TT

By Table 4.9, C has dimension 17 and therefore M is conjugate to a subgroup of H by Lemma 5.24, and C has one of the following types: G_2A_1, G_2T_3, $B_2A_1A_1T_1$, $B_2A_1T_4$, $A_2A_2T_1$, $A_2A_1A_1A_1$, $A_2A_1A_1T_3$, $A_1^5T_2$. By Table 5.36, Remark 5.10 and Lemma 5.4, there is only one H-class of SL(2, 5) subgroups with this fusion pattern in H, namely, the class which corresponds to $6^2 + 1^4$, and therefore only one G-class of such subgroups in G. Moreover, by Lemma 5.35, C can have rank at most 3 so has type G_2A_1.

Case 2294: 2A, 3D, 4G, 5H, 6S, 10CCC

By Table 4.9, C has dimension 133 and therefore M is conjugate to a subgroup of H by Lemma 5.24, and C has type E_7. By Table 5.36, Remark 5.10

and Lemma 5.4, there is only one H-class of SL(2, 5) subgroups with this fusion pattern in H, namely, the class which corresponds to $2^2 + 1^{12}$. There is therefore only one G-class of such subgroups in G.

Cases 2305, 2324, 2342, 2900, 2918, 2937, 3500:

In each of these cases there is an SL(2, 5)-subgroup with central involution z. Let $x \neq z$ be an involution of H of type 2B. Then $\exists g \in G$ such that $x^g = z$. If M is an SL(2, 5)-subgroup of H with central involution x, then M^g is an SL(2, 5)-subgroup of H with central involution z. Hence M^g corresponds to a character with no trivial constituents, and there is therefore an element of type 3A which centralizes M^g. But then M is conjugate to a subgroup of $\mathcal{A}$, and since there is only one $\mathcal{A}$-class of SL(2, 5)-subgroups of $\mathcal{A}$ with each of these fusion patterns, there is only one G-class of SL(2, 5)-subgroups of G with each of these fusion patterns.

Cases 2458, 2475, 2476, 2491, 2493, 2511, 3052, 3063, 3069, 3088, 3089, 3105, 3141, 3628, 3645, 3665, 3717, 3847, 3868, 4438:

Since in each of these cases, M is conjugate to a subgroup of H with central involution z, the number of G-classes of subgroups with each fusion pattern is the number of H-classes which, in turn, is the number of occurrences of each fusion pattern in Table 5.14.

Theorem 8.1 Suppose M is an SL(2, 5)-subgroup of G. Then M is conjugate to every other SL(2, 5)-subgroup of G with the same fusion pattern unless M has fusion pattern 174, 188, 633, 786, 2476 or 2493. There are two G-classes of SL(2, 5)-subgroups with fusion patterns 786, 2476 and 2493 respectively. All fusion patterns are listed in Table 8.2 along with their centralizer types.

Table 8.2 Conjugacy classes of SL(2, 5) subgroups of G.

These are the fusion patterns of SL(2, 5) subgroups which are contained in G (= E_8), some important subgroups in which the SL(2, 5) subgroups are contained, their connected centralizers, and the number of conjugacy classes in each in G.

Fusion pattern	Subgroups of G which contain SL(2, 5) subgroups of the given fusion pattern	Connected centralizer type	Cases in Table 5.36 which represent this fusion pattern	Number of conjugacy classes in G
3 2A, 3A, 4A, 5A, 6C, 10N	$\mathcal{A}, \Delta, \Omega, H$	B_2T_1 or A_2A_1	31, 42	1
19 2A, 3A, 4A, 5B, 6C, 10D	$\mathcal{A}, \Delta, \Omega, H$	A_2A_1	29	1
22 2A, 3A, 4A, 5B, 6C, 10Z	$\mathcal{A}, \Delta, \Omega, H$	$A_1A_1T_1$	30, 36, 40	1
37 2A, 3A, 4A, 5C, 6C, 10U	$\mathcal{A}, \Delta, \Omega, H$	$A_1A_1A_1$ or A_2T_1	32, 34	1
57 2A, 3A, 4A, 5D, 6C, 10HH	Ω, H	A_1A_1	35, 41	1
152 2A, 3A, 4D, 5A, 6G, 10L	Δ, H	$A_1A_1A_1$	27, 82	1
170 2A, 3A, 4D, 5B, 6G, 10P	Δ, H	$A_1A_1A_1$	26, 80	1
174 2A, 3A, 4D, 5B, 6G, 10JJ	Δ, H	A_1	78	≥ 1**
188 2A, 3A, 4D, 5C, 6G, 10KK	Δ, H	A_1	79	≥ 1**
210 2A, 3A, 4D, 5D, 6G, 10AAA	H	A_1A_1	81, 28	1
598 2A, 3B, 4A, 5B, 6F, 10Z	$\mathcal{A}, H$	A_1T_1	39, 43	1
613 2A, 3B, 4A, 5C, 6F, 10U	$\mathcal{A}, H$	A_1A_1	37	1

633 2A, 3B, 4A, 5D, 6F, 10HH	Я, H	A_1	44, 38	$\geq 1^*$
750 2A, 3B, 4D, 5B, 6O, 10JJ	Я, H	A_1A_1	48	1
764 2A, 3B, 4D, 5C, 6O, 10KK	Я, H	A_1A_1	46	1
785 2A, 3B, 4D, 5D, 6O, 10ZZ	Я, Δ, H	G_2A_1	45	1
786 2A, 3B, 4D, 5D, 6O, 10AAA	Я, Δ, H	$A_1A_1A_1$ or A_2T_1	33, 47	2
800 2A, 3B, 4D, 5E, 6O, 10WW	Я, Δ, H	G_2A_1	49	1
934 2A, 3B, 4E, 5D, 6P, 10YY	Я, Δ, Ω, H	B_4A_1	8, 17, 18	1
951 2A, 3B, 4E, 5E, 6P, 10BBB	Я, Δ, Ω, H	F_4A_1	7, 16	1
1310 2A, 3C, 4D, 5A, 6R, 10L	Ω, H	B_3	24, 84, 88	1
1328 2A, 3C, 4D, 5B, 6R, 10P	Ω, H	B_3	23, 86	1
1368 2A, 3C, 4D, 5D, 6R, 10AAA	Я, Ω, H	A_3A_1	21, 25, 87	1
1401 2A, 3C, 4D, 5F, 6R, 10EEE	Я, Δ, Ω, H	A_5 or B_3G_2	22, 85	1
1419 2A, 3C, 4D, 5G, 6R, 10DDD	Я, Δ, Ω, H	A_5 or B_3G_2	20, 83	1
1504 2A, 3C, 4E, 5D, 6L, 10YY	Я, H	B_3A_1	6, 19	1
1556 2A, 3C, 4E, 5G, 6L, 10TT	Я, H	G_2A_1	5	1
2294 2A, 3D, 4G, 5H, 6S, 10CCC	Я, Δ, Ω, H	E_7	1	1
2305 2B, 3A, 4B, 5A, 6A, 10A	Я, Δ, Ω, H	B_4 or C_4	13, 58	1
2324 2B, 3A, 4B, 5B, 6A, 10O	Я, Δ, Ω, H	A_4, B_3A_1 or B_2G_2	14, 60, 70	1
2342 2B, 3A, 4B, 5C, 6A, 10FF	Я, Δ, Ω, H	B_2B_2	15, 59, 71	1
2458 2B, 3A, 4C, 5A, 6I, 10EE	Δ, H	A_1A_1	None: see Table 5.14	1
2475 2B, 3A, 4C, 5B, 6I, 10AA	Δ, H	A_1A_1	None: see Table 5.14	1
2476 2B, 3A, 4C, 5B, 6I, 10II	Я, Δ, H	A_1T_1	57	2

2491 2B, 3A, 4C, 5C, 6I, 10T	Δ, H	B_2	None: see Table 5.14	1
2493 2B, 3A, 4C, 5C, 6I, 10OO	$\mathcal{A}$, Δ, H	T_2	None: see Table 5.14	2
2511 2B, 3A, 4C, 5D, 6I, 10MM	H	A_1	None: see Table 5.14	1
2900 2B, 3B, 4B, 5B, 6H, 10O	$\mathcal{A}$, H	$B_2 A_1$	11, 50, 61	1
2918 2B, 3B, 4B, 5C, 6H, 10FF	$\mathcal{A}$, H	$A_1 A_1 A_1$	12, 51, 72	1
2937 2B, 3B, 4B, 5D, 6H, 10MM	$\mathcal{A}$, H	$A_1 A_1$	9	1
3052 2B, 3B, 4C, 5B, 6K, 10II	$\mathcal{A}$, Ω, H	$B_2 T_1$	63, 75	1
3063 2B, 3B, 4C, 5C, 6H, 10OO	$\mathcal{A}$, H	A_1	74?	1
3069 2B, 3B, 4C, 5C, 6K, 10OO	$\mathcal{A}$, Ω, H	$A_1 A_1 A_1$	74?, 76?	1
3088 2B, 3B, 4C, 5D, 6K, 10PP	$\mathcal{A}$, Δ, Ω, H	$B_3 A_1$	62	1
3089 2B, 3B, 4C, 5D, 6K, 10QQ	$\mathcal{A}$, Δ, Ω, H	$A_3 T_1$	65, 73	1
3105 2B, 3B, 4C, 5E, 6K, 10XX	$\mathcal{A}$, Δ, Ω, H	$B_3 T_1$	64, 67	1
3141 2B, 3B, 4C, 5G, 6K, 10SS	Ω, H	A_3	66, 76?	1
3500 2B, 3C, 4B, 5C, 6J, 10FF	$\mathcal{A}$, H	B_2	10, 52	1
3628 2B, 3C, 4C, 5B, 6J, 10II	H	$A_1 A_1$	55, 77	1
3645 2B, 3C, 4C, 5C, 6J, 10OO	$\mathcal{A}$, H	$A_1 T_1$	54, 56?	1
3665 2B, 3C, 4C, 5D, 6J, 10QQ	$\mathcal{A}$, H	$B_2 T_1$	53, 68	1
3717 2B, 3C, 4C, 5G, 6J, 10SS	H	B_2	56?, 69	1
3847 2B, 3C, 4F, 5F, 6Q, 10B	$\mathcal{A}$, Δ, Ω, H	B_6	3	1
3868 2B, 3C, 4F, 5G, 6Q, 10FFF	$\mathcal{A}$, Δ, Ω, H	D_6	4	1
4438 2B, 3D, 4F, 5G, 6M, 10FFF	$\mathcal{A}$, H	B_5	2	1

*We actually know more than this. See discussion of Alt_5 case 844 (p. 124).

**These groups are conjugate to subgroups of $C(\mathcal{E}) \circ \mathcal{E}$. See Remark 8.0 (p. 140).

Remark 8.3 It is clear that for each conjugacy class K of SL(2, 5) subgroups which have nonrational elements of order 10, that there are two nonconjugate classes of embeddings of SL(2, 5) subgroups of K. As for those classes which have rational elements of order 10, all of them have representatives in H. Suppose K is an SL(2, 5) conjugacy class where the elements of K have rational elements of order 10. Then by Theorem 8.1, K consists of all SL(2, 5) subgroups of G with the given fusion pattern (except in case 2493), so if there is any element of K afforded by an embedding of SL(2, 5) into H which is unchanged by an outer twist and if there is an element of H which induces that outer twist, then there is only one class of embeddings affording elements of K. Such an element exists in H if any of the irreducible constituents are odd. We produce such an embedding for each rational class K. In case 2493, we produce two embeddings since there are two classes of SL(2, 5) subgroups with fusion pattern 2493.

1556: $6^2 + 1^4$	3141: $4 + 3 + 3' + 1^6$
2342: $2^2 + 2'^2 + 4 + 1^4$	3500: $4_f^2 + 5 + 1^3$
2493: $5^2 + 4 + 1^2$, $4 + 3^2 + 3'^2$	3645: $5 + 4^2 + 1^3$
(these embeddings are nonconjugate)	
2918: $4_f^2 + 4 + 1^4$	3717: $5 + 3 + 3' + 1^5$
3063: $5^3 + 1$	3868: $4 + 1^{12}$
3069: $4^3 + 1^4$	4438: $5 + 1^{11}$

Appendix

Although not used in the main body of the paper, the following Lemmas are interesting in their own right. In what follows, we classify up to conjugacy all nines-groups of G which do not contain elements of type 3A.

Lemma 9.1 Suppose E is a 9-subgroup of G. If E is 3CD-pure then E contains 4 elements of type 3C and 4 elements of type 3D. Moreover, if E is such a group, E is unique up to conjugacy and C(E) has type D_6T_2.

Proof. Let E be a 3CD-pure 9-subgroup of G. Then the dimension of C(E) is one of 40, 54, 68, 82 or 96 by the orthogonality relations. Let x be an element of type 3D in E. Then C(x) = XS where X has type E_7, and S is a one dimensional torus, and x ε S since $|Z(X)| = 2$. Now $\exists y \in X \cap E\backslash<x>$ with E = <x, y> so C(E) = $C_X(y)$ o S. By Table 1.19, since E is 3CD-pure, y has type 3D[S] or 3E[S] so $C_X(y)$ has type $A_1D_5T_1$ which has dimension 49 or type D_6T_1 which has dimension 67 and C(E) has dimension 50 or 68. Only dimension 68 is on the list of possibilities however, so y has type 3D[S], E is unique up to conjugacy having 4 elements of each type, and C(E) has type D_6T_2.

The only remaining case is when E is 3C-pure. In this case if x ε E, C(x) = XS where X has type D_7, and S is a one dimensional torus. Now $\exists y \in X \cap E\backslash<x>$ such that E = <x, y> and C(E) = $C_X(y)$ o S. Now $C_X(y)$ has dimension 39 and the possibilities are: A_5T_2 (dim 37), D_6T_1 (dim 67), A_6T_1 (dim 49), $D_5A_1T_1$ (dim 49), $D_4A_2T_1$ (dim 37), $A_3A_3T_1$ (dim 31) and $A_4A_1A_1T_1$ (dim 31). None of these has the correct dimension. ∎

Lemma 9.2 Suppose E is a BD-pure 9-subgroup of G. Then either E is 3B-pure, or E has 2 elements of type 3B, and 6 elements of type 3D. In case E is

3B-pure, its centralizer has type one of A_2^4 or D_4T_4. If E has the other type, then its centralizer has type E_6T_2. Moreover, these nines-groups are unique up to conjugacy.

Proof. Let r be the number of elements of E of type 3B, s the number of elements of E of type 3D. Since every element of G is conjugate to its inverse, r and s are even. If $s \neq 0$, let x be an element of E of type 3D. Then $C(x) = XS$ where X has type E_7, and S is a one dimensional torus, and $x \in S$ since $|Z(X)| = 2$. Now $\exists y \in X \cap E \backslash <x>$ with $E = <x, y>$ so $C(E) = C_X(y)$ o S. We now study each case individually.

<u>Case r=0, s=8</u>: By the orthogonality relations, the dimension of C(E) is 96. But by Table 1.19, there are no centralizers of elements of order 3 in X of dimension 95, so there are no 3D-pure nines-groups in G.

<u>Case r=2, s=6</u>: By the orthogonality relations, the dimension of C(E) is 80. By Table 1.19, only elements of type 3B[s] have centralizers of dimension 79, so there is only one G-class of nines-groups with this fusion pattern, and C(E) has type E_6T_2.

<u>Case r=4, s=4</u>: By the orthogonality relations, the dimension of C(E) is 64. But by Table 1.19, there are no centralizers of elements of order 3 in X of dimension 63, so there are no nines-groups of this type in G.

<u>Case r=6, s=2</u>: By the orthogonality relations, the dimension of C(E) is 48. But by Table 1.19, there are no centralizers of elements of order 3 in X of dimension 47, so there are no nines-groups of this type in G.

<u>Case r=8, s=0</u>: By [Gr '91](11.4), C(E) has type A_2^4 or D_4T_4, and in each case E is unique up to conjugacy.

This exhaustion of cases gives us the result.∎

Lemma 9.3 Suppose E is a nines-group of G with elements of type 3B, 3C and 3D, and no elements of type 3A. Then E contains 4 elements of type 3B, 2 elements of type 3C and 2 elements of type 3D, and C(E) has type $A_1 D_5 T_2$. Moreover, E is unique up to conjugacy.

Proof. Let r be the number of elements of E of type 3B, t the number of elements of E of type 3C, and s the number of elements of E of type 3D. Since every element of G is conjugate to its inverse, r, t and s are even. Let x be an element of E of type 3D. Then C(x) = XS where X has type E_7, and S is a one dimensional torus, and $x \in S$ since $|Z(X)| = 2$. Now $\exists y \in X \cap E \setminus \langle x \rangle$ with $E = \langle x, y \rangle$ so $C(E) = C_X(y) \circ S$. We now study each case individually.

<u>Case r=2, t=2, s=4</u>: By the orthogonality relations, the dimension of C(E) is 66. But by Table 1.19, there are no centralizers of elements of order 3 in X of dimension 65, so there are no nines-groups of this type in G.

<u>Case r=2, t=4, s=2</u>: By the orthogonality relations, the dimension of C(E) is 52. But by Table 1.19, there are no centralizers of elements of order 3 in X of dimension 47, so there are no nines-groups of this type in G.

<u>Case r=4, t=2, s=2</u>: By the orthogonality relations, the dimension of C(E) is 50. Now by Table 1.19, there are two classes of elements of order 3 in X with centralizer of dimension 49, namely, 3A[X], and 3D[X]. But 3A[X] $\subseteq$ 3A[G] and we are assuming that E doesn't contain any elements of type 3A[G], so the only possibility is that the elements of E of order 3 have type 3D[X], so E is unique up to conjugacy, and C(E) has type $A_1 D_5 T_2$.

Since we have exhausted all possibilities, the result is proved. ∎

Table 9.4 Centralizers of elements of order three in a group of type D_7:

Centralizer type	Dimension	Multiplicity
$A_5 T_2$	37	4
$A_6 T_1$	49	4

D_6T_1	67	2
$D_5A_1T_1$	39	4
$D_4A_2T_1$	37	4
$A_3A_3T_1$	31	4
$A_4A_1A_1T_1$	31	4

Lemma 9.5 Suppose E is a BC-pure 9-subgroup of G. Then either E is 3B-pure, or E has 2 elements of type 3B, and 6 elements of type 3C. In case E is 3B-pure, its centralizer has type one of A_2^4 or D_4T_4. If E has the other type, then its connected centralizer has type A_5T_3 or $A_2D_4T_2$, and in each case E is unique up to conjugacy, (that is, there are four classes of BC-pure nines-groups in G).

Proof. Let r be the number of elements of E of type 3B, t the number of elements of E of type 3C. Since every element of G is conjugate to its inverse, r and s are even. If $t \neq 0$, let x be an element of E of type 3C. Then $C(x) = XS$ where X has type D_7, and S is a one dimensional torus, and $x \in S$ since $|Z(X)| = 4$. Now $\exists y \in X \cap E \backslash <x>$ with $E = <x, y>$ so $C(E) = C_X(y)$ o S. We now study each case individually.

<u>Case r=0, s=8</u>: By Lemma 9.1 there are no 3C-pure nines-groups in G.

<u>Case r=2, s=6</u>: By the orthogonality relations, the dimension of C(E) is 38. There are two connected centralizer types of elements of order 3 in X with dimension 37, namely type A_5T_2, and $A_2D_4T_1$. Hence there are two classes of nines-group with this fusion pattern and C(E) has type A_5T_3 in one case and $A_2D_4T_2$ in the other.

<u>Case r=4, s=4</u>: By the orthogonality relations, the dimension of C(E) is 36. But there are no centralizers of elements of order 3 in X of dimension 35, so there are no nines-groups of this type in G.

Case r=6, s=2: By the orthogonality relations, the dimension of C(E) is 34. But there are no centralizers of elements of order 3 in X of dimension 33, so there are no nines-groups of this type in G.

Case r=8, s=0: By [Gr '91](11.4), C(E) has type A_2^4 or $D_4 T_4$, and in each case E is unique up to conjugacy.

This exhaustion of cases gives us the result. ∎

Table of Notation and Definitons

Notation or definition	Place defined
central product	Definition 1.4
perfect group, quasisimple group	Definition 1.7
G	Notation 1.12
H	Notation 1.12
$\mathcal{S}$	Notation 1.12
$\mathcal{E}$	Notation 1.12
$\mathcal{F}$	Notation 1.12
$\mathcal{G}$	Notation 1.12
$\mathcal{A}$	Notation 1.12
$\mathcal{D}$	Notation 1.12
Δ	Notation 1.9
Ω	Notation 1.9
θ	Notation 1.9
$\mathcal{T}$	Notation 1.12
χ	Notation 1.9
fours-group, nines-group, sixteens-group	Note 1.13
rational element	Definition 1.17
ξ	Notation 1.18
κ	Notation 1.18
quasiprojection	Definition 1.28
fusion pattern	Definition 4.1
4_f	Notation 4.10
outer twist	Remark 4.11
Hamming code (extended)	Definition 4.20
tetracode	Definition 4.20
$\tilde{H}$	Notation 5.2
K	Notation 5.2
z	Notation 5.2
z_1	Notation 5.2
multiplicity	Definition 5.31
ZDC	Notation 7.3

References

[Artin '57] Artin, E., *Geometric Algebra*, Interscience Publishers Inc., New York, 1957.

[Atlas '85] Conway, J. H., Curtis, R. T., Norton, S. P., Parker, R. P., Wilson, R. A., *Atlas of finite groups*, Clarendon Press, Oxford, 1985.

[Carter '89] Carter, R., *Simple groups of Lie type*. J. Wiley and Sons Ltd, New York, 1989.

[CGL] Cohen, A.M., Griess, R., Jr., Lisser, Bert, 'The group L(2, 61) embeds in the Lie group of type E_8', *Comm. in Algebra,*, **21**(6), (1993) 1889-1907.

[CoGr '87] Cohen, A. M. and Griess, R., Jr, 'On finite simple subgroups of the complex Lie group of type E_8', *Proc. Symp. Pure Math.* **47** (1987), 367-405.

[CoGr '93] Cohen, A. M. and Griess, R., Jr, 'Non-local Lie primitive subgroups of Lie groups', *Canadian Journal of Mathematics* **45** (1993), 88-103.

[CoWa '83] Cohen, A. M. and Wales, D. B., 'Finite subgroups of $G_2(\mathbb{C})$', *Comm. Algebra*, **11** (1983), 441-459.

[CoWa '92] Cohen, A. M. and Wales, D. B., 'Finite subgroups of $F_4(\mathbb{C})$ and $E_6(\mathbb{C})$', preprint.

[FuHa '91] Fulton, W. and Harris, J., *Representation Theory: A First Course*, Springer-Verlag, New York, 1991.

[Gr '91] Griess, R., Jr, 'Elementary abelian p-subgroups of algebraic groups', *Geometriae Dedicata* **39** (1991), 253-305.

[Gr unp.] Griess, R., Jr., 'Sporadic groups', preprint.

[GrRy '96] Griess, R., Jr., Ryba, A. J. E., 'Embeddings of $PGL_2(31)$ and $SL_2(32)$ in $E_8(\mathbb{C})$', preprint.

[Hall '59] Hall, M., Jr., *The theory of groups*, The MacMillan Co., New York, 1959.

[Hup '67] Huppert, B., *Endliche Gruppen I*, Springer-Verlag, Berlin, 1967.

[Kac '90] Kac, V.G., *Infinite Dimensional Lie Algebras*, Cambridge University Press, Cambridge, 1990.

[St '81] Steinberg, R., 'Generators, relations and coverings of algebraic groups', *J. Algebra* II **71** (1981), 521-543.

[Serre '96] Serre, J. P., 'Exemples de plongements des groupes $PSL(2, \mathbf{F}_p)$ dans des groupes de Lie simples, *Inventiones Math.* **124** (1996), 525-562.

[Tits '55] Tits, J., 'Sous-algè bres des Algè bres de Lie Semi-simples', *Seminaire Bourbaki, Textes Des Confé rences, Exposé s 101 à 119*, 7(1954/1955), 119-1 - 119-18.

[Wood '89] Wood, J. A., 'Spinor groups and algebraic coding theory', *J. Combin. Theory* (1989), 277-313.

Darrin D. Frey
Department of Science and Mathematics
Cedarville College
P.O. Box 601
Cedarville, Ohio 45314
freyd@cedarville.edu

Editorial Information

To be published in the *Memoirs*, a paper must be correct, new, nontrivial, and significant. Further, it must be well written and of interest to a substantial number of mathematicians. Piecemeal results, such as an inconclusive step toward an unproved major theorem or a minor variation on a known result, are in general not acceptable for publication. *Transactions* Editors shall solicit and encourage publication of worthy papers. Papers appearing in *Memoirs* are generally longer than those appearing in *Transactions* with which it shares an editorial committee.

As of January 31, 1998, the backlog for this journal was approximately 9 volumes. This estimate is the result of dividing the number of manuscripts for this journal in the Providence office that have not yet gone to the printer on the above date by the average number of monographs per volume over the previous twelve months, reduced by the number of issues published in four months (the time necessary for preparing an issue for the printer). (There are 6 volumes per year, each containing at least 4 numbers.)

A Copyright Transfer Agreement is required before a paper will be published in this journal. By submitting a paper to this journal, authors certify that the manuscript has not been submitted to nor is it under consideration for publication by another journal, conference proceedings, or similar publication.

Information for Authors and Editors

Memoirs are printed by photo-offset from camera copy fully prepared by the author. This means that the finished book will look exactly like the copy submitted.

The paper must contain a *descriptive title* and an *abstract* that summarizes the article in language suitable for workers in the general field (algebra, analysis, etc.). The *descriptive title* should be short, but informative; useless or vague phrases such as "some remarks about" or "concerning" should be avoided. The *abstract* should be at least one complete sentence, and at most 300 words. Included with the footnotes to the paper, there should be the 1991 *Mathematics Subject Classification* representing the primary and secondary subjects of the article. This may be followed by a list of *key words and phrases* describing the subject matter of the article and taken from it. A list of the numbers may be found in the annual index of *Mathematical Reviews*, published with the December issue starting in 1990, as well as from the electronic service e-MATH [**telnet e-MATH.ams.org** (or **telnet 130.44.1.100**). Login and password are **e-math**]. For journal abbreviations used in bibliographies, see the list of serials in the latest *Mathematical Reviews* annual index. When the manuscript is submitted, authors should supply the editor with electronic addresses if available. These will be printed after the postal address at the end of each article.

Electronically prepared papers. The AMS encourages submission of electronically prepared papers in $\mathcal{A}_{\mathcal{M}}\mathcal{S}$-TeX or $\mathcal{A}_{\mathcal{M}}\mathcal{S}$-LaTeX. The Society has prepared author packages for each AMS publication. Author packages include instructions for preparing electronic papers, the *AMS Author Handbook*, samples, and a style file that generates the particular design specifications of that publication series for both $\mathcal{A}_{\mathcal{M}}\mathcal{S}$-TeX and $\mathcal{A}_{\mathcal{M}}\mathcal{S}$-LaTeX.

Authors with FTP access may retrieve an author package from the Society's Internet node **e-MATH.ams.org** (130.44.1.100). For those without FTP access, the author package can be obtained free of charge by sending e-mail to **pub@ams.org** (Internet) or from the Publication Division, American Mathematical Society, P.O. Box 6248, Providence, RI 02940-6248. When requesting an author package, please specify $\mathcal{AMS}$-TEX or $\mathcal{AMS}$-LATEX, Macintosh or IBM (3.5) format, and the publication in which your paper will appear. Please be sure to include your complete mailing address.

Submission of electronic files. At the time of submission, the source file(s) should be sent to the Providence office (this includes any TEX source file, any graphics files, and the DVI or PostScript file).

Before sending the source file, be sure you have proofread your paper carefully. The files you send must be the EXACT files used to generate the proof copy that was accepted for publication. For all publications, authors are required to send a printed copy of their paper, which exactly matches the copy approved for publication, along with any graphics that will appear in the paper.

TEX files may be submitted by email, FTP, or on diskette. The DVI file(s) and PostScript files should be submitted only by FTP or on diskette unless they are encoded properly to submit through e-mail. (DVI files are binary and PostScript files tend to be very large.)

Files sent by electronic mail should be addressed to the Internet address **pub-submit@ams.org**. The subject line of the message should include the publication code to identify it as a Memoir. TEX source files, DVI files, and PostScript files can be transferred over the Internet by FTP to the Internet node **e-math.ams.org** (130.44.1.100).

Electronic graphics. Figures may be submitted to the AMS in an electronic format. The AMS recommends that graphics created electronically be saved in Encapsulated PostScript (EPS) format. This includes graphics originated via a graphics application as well as scanned photographs or other computer-generated images.

If the graphics package used does not support EPS output, the graphics file should be saved in one of the standard graphics formats—such as TIFF, PICT, GIF, etc.—rather than in an application-dependent format. Graphics files submitted in an application-dependent format are not likely to be used. No matter what method was used to produce the graphic, it is necessary to provide a paper copy to the AMS.

Authors using graphics packages for the creation of electronic art should also avoid the use of any lines thinner than 0.5 points in width. Many graphics packages allow the user to specify a "hairline" for a very thin line. Hairlines often look acceptable when proofed on a typical laser printer. However, when produced on a high-resolution laser imagesetter, hairlines become nearly invisible and will be lost entirely in the final printing process.

Screens should be set to values between 15% and 85%. Screens which fall outside of this range are too light or too dark to print correctly.

Any inquiries concerning a paper that has been accepted for publication should be sent directly to the Editorial Department, American Mathematical Society, P. O. Box 6248, Providence, RI 02940-6248.

Editors

This journal is designed particularly for long research papers (and groups of cognate papers) in pure and applied mathematics. Papers intended for publication in the *Memoirs* should be addressed to one of the following editors:

Ordinary differential equations, partial differential equations, and applied mathematics to JOHN MALLET-PARET, Division of Applied Mathematics, Brown University, Providence, RI 02912-9000; electronic mail: `jmp@cfm.brown.edu`.

Harmonic analysis, representation theory, and Lie theory to ROBERT J. STANTON, Department of Mathematics, The Ohio State University, 231 West 18th Avenue, Columbus, OH 43210-1174; electronic mail: `stanton@math.ohio-state.edu`.

Ergodic theory and dynamical systems to ROBERT F. WILLIAMS, Department of Mathematics, University of Texas at Austin, Austin, TX 78712-1082; e-mail: `bob@math.utexas.edu`

Real and harmonic analysis and geometric partial differential equations to WILLIAM BECKNER, Department of Mathematics, University of Texas at Austin, Austin, TX 78712-1082; e-mail: `beckner@math.utexas.edu`.

Algebra to CHARLES CURTIS, Department of Mathematics, University of Oregon, Eugene, OR 97403-1222 e-mail: `curtis@bright.uoregon.edu`

Algebraic topology and cohomology of groups to STEWART PRIDDY, Department of Mathematics, Northwestern University, 2033 Sheridan Road, Evanston, IL 60208-2730; e-mail: `s_priddy@math.nwu.edu`.

Differential geometry and global analysis to CHUU-LIAN TERNG, Department of Mathematics, Northeastern University, Huntington Avenue, Boston, MA 02115-5096; e-mail: `terng@neu.edu`.

Probability and statistics to RODRIGO BAÑUELOS, Department of Mathematics, Purdue University, West Lafayette, IN 47907-1968; e-mail: `banuelos@math.purdue.edu`.

Combinatorics and Lie theory to PHILIP J. HANLON, Department of Mathematics, University of Michigan, Ann Arbor, MI 48109-1003; e-mail: `hanlon@math.lsa.umich.edu`.

Logic to THEODORE SLAMAN, Department of Mathematics, University of California at Berkeley, Berkeley, CA 94720-3840; e-mail: `slaman@math.berkeley.edu`.

Number theory and arithmetic algebraic geometry to ALICE SILVERBERG, Department of Mathematics, Harvard University, 1 Oxford St.–Science Center, Cambridge, MA 02138; e-mail: `silver@math.ohio-state.edu`.

Complex analysis and complex geometry to DANIEL M. BURNS, Department of Mathematics, University of Michigan, Ann Arbor, MI 48109-1003; e-mail: `dburns@math.lsa.umich.edu`.

Algebraic geometry and commutative algebra to LAWRENCE EIN, Department of Mathematics, University of Illinois, 851 S. Morgan (M/C 249), Chicago, IL 60607-7045; e-mail: `ein@uic.edu`.

Geometric topology, knot theory, hyperbolic geometry, and general topoogy to JOHN LUECKE, Department of Mathematics, University of Texas at Austin, Austin, TX 78712-1082; e-mail: `luecke@math.utexas.edu`.

Partial differential equations and applied mathematics to BARBARA LEE KEYFITZ, Department of Mathematics, University of Houston, 4800 Calhoun, Houston, TX 77204-3476; e-mail: `keyfitz@uh.edu`

All other communications to the editors should be addressed to the Managing Editor, PETER SHALEN, Department of Mathematics, University of Illinois, 851 S. Morgan (M/C 249), Chicago, IL 60607-7045; e-mail: `shalen@math.uic.edu`.

Selected Titles in This Series

(Continued from the front of this publication)

605 **Liangqing Li,** Classification of simple C^*-algebras: Inductive limits of matrix algebras over trees, 1997

604 **Hajnal Andréka, Steven Givant, and István Némethi,** Decision problems for equational theories of relation algebras, 1997

603 **Bruce N. Allison, Saeid Azam, Stephen Berman, Yun Gao, and Arturo Pianzola,** Extended affine Lie algebras and their root systems, 1997

602 **Igor Fulman,** Crossed products of von Neumann algebras by equivalence relations and their subalgebras, 1997

601 **Jack E. Graver and Mark E. Watkins,** Locally finite, planar, edge-transitive graphs, 1997

600 **Ambar Sengupta,** Gauge theory on compact surfaces, 1997

599 **Tai-Ping Liu and Yanni Zeng,** Large time behavior of solutions for general quasilinear hyperbolic-parabolic systems of conservation laws, 1997

598 **Valentina Barucci, David E. Dobbs, and Marco Fontana,** Maximality properties in numerical semigroups and applications to one-dimensional analytically irreducible local domains, 1997

597 **Ragnar-Olaf Buchweitz and John J. Millson,** CR-geometry and deformations of isolated singularities, 1997

596 **Paul S. Bourdon and Joel H. Shapiro,** Cyclic phenomena for composition operators, 1997

595 **Eldar Straume,** Compact connected Lie transformation groups on spheres with low cohomogeneity, II, 1997

594 **Solomon Friedberg and Hervé Jacquet,** The fundamental lemma for the Shalika subgroup of $GL(4)$, 1996

5^ **Ajit Iqbal Singh,** Completely positive hypergroup actions, 1996

592 **P. Kirk and E. Klassen,** Analytic deformations of the spectrum of a family of Dirac operators on an odd-dimensional manifold with boundary, 1996

591 **Edward Cline, Brian Parshall, and Leonard Scott,** Stratifying endomorphism algebras, 1996

590 **Chris Jantzen,** Degenerate principal series for symplectic and odd-orthogonal groups, 1996

589 **James Damon,** Higher multiplicities and almost free divisors and complete intersections, 1996

588 **Dihua Jiang,** Degree 16 Standard L-function of $GSp(2) \times GSp(2)$, 1996

587 **Stéphane Jaffard and Yves Meyer,** Wavelet methods for pointwise regularity and local oscillations of functions, 1996

586 **Siegfried Echterhoff,** Crossed products with continuous trace, 1996

585 **Gilles Pisier,** The operator Hilbert space OH, complex interpolation and tensor norms, 1996

584 **Wayne W. Barrett, Charles R. Johnson, and Raphael Loewy,** The real positive definite completion problem: Cycle completability, 1996

583 **Jin Nakagawa,** Orders of a quartic field, 1996

582 **Darryl McCollough and Andy Miller,** Symmetric automorphisms of free products, 1996

581 **Martin U. Schmidt,** Integrable systems and Riemann surfaces of infinite genus, 1996

580 **Martin W. Liebeck and Gary M. Seitz,** Reductive subgroups of exceptional algebraic groups, 1996

579 **Samuel Kaplan,** Lebesgue theory in the bidual of $C(X)$, 1996

578 **Ale Jan Homburg,** Global aspects of homoclinic bifurcations of vector fields, 1996

577 **Freddy Dumortier and Robert Roussarie,** Canard cycles and center manifolds, 1996

(See the AMS catalog for earlier titles)